AN ANALYTICAL CALCULUS

VOLUME III

AN
ANALYTICAL CALCULUS
FOR SCHOOL AND UNIVERSITY

by

E. A. MAXWELL
Fellow of Queens' College, Cambridge

VOLUME III

CAMBRIDGE
AT THE UNIVERSITY PRESS
1966

CAMBRIDGE UNIVERSITY PRESS
Cambridge, New York, Melbourne, Madrid, Cape Town, Singapore, São Paulo, Delhi

Cambridge University Press
The Edinburgh Building, Cambridge CB2 8RU, UK

Published in the United States of America by Cambridge University Press, New York

www.cambridge.org
Information on this title: www.cambridge.org/9780521056984

© Cambridge University Press 1954

First published 1954
Reprinted 1966
This digitally printed version 2008

A catalogue record for this publication is available from the British Library

ISBN 978-0-521-05698-4 hardback
ISBN 978-0-521-09037-7 paperback

CONTENTS

INTRODUCTORY NOTE: COORDINATE SYSTEMS
IN SPACE OF THREE DIMENSIONS

PAGE 1

CHAPTER XIV: PARTIAL DIFFERENTIATION

CHAPTER XV: MAXIMA AND MINIMA

CHAPTER XVI: JACOBIANS

CHAPTER XVII: MULTIPLE INTEGRALS

CHAPTER XVIII: THE SKETCHING OF CURVES

PREFACE

This volume continues for two or more variables the work under-
taken for one in the preceding volumes. Though it seems wise to
relax standards somewhat, the aim is still reasonable rigour; for
example, the formula $\iint f(r, \theta)\, r\, dr\, d\theta$ for a double integral in
polar coordinates and the formula $\iiint f(r, \theta, \phi)\, r^2 \sin\theta\, dr\, d\theta\, d\phi$ for
a triple integral in spherical polar coordinates are derived by
methods intended to provide a convincing proof without recourse
to the usual 'small elements' of apparently approximately rele-
vant values. I hope that those who pass from this book to the
greater exactitude of analysis will be able to do so with some
appreciation of the nature of analytical argument.

I have allowed myself the pleasure of a chapter on the sketching
of curves, which is rather out of fashion at the moment. There is
no better field for developing that interplay of intuition and logic
which is half the thrill of mathematics, and a revival, kept
within bounds, is to be desired. As far as I know, the treatment
(though foreshadowed by Salmon) has not been developed in
such detail in text-books at this level.

I must again express my thanks to Dr Edmonds and Dr Cassels
for their valuable help, and to pupils who have assisted me in
the checking of answers. My gratitude is also due to the staff of
the Cambridge University Press for their care and attention.

E. A. M.

QUEENS' COLLEGE, CAMBRIDGE
March, 1954

Opportunity has been taken while re-printing to make a few
changes and corrections. These are usually slight, but where use is
being made of this and the earlier printings together, attention
should be drawn to pages 92, 154, 176, 179.

E. A. M.

November, 1959

INTRODUCTORY NOTE

COORDINATE SYSTEMS IN SPACE OF
THREE DIMENSIONS

The object of this note is to explain briefly how some of the common coordinate systems are defined in space of three dimensions. For subsequent developments, reference must, of course, be made to a text-book of analytical solid geometry.

Let O be a point fixed in space, and Oz a fixed line through it (fig. 104). For convenience of reference, we shall allude to Oz as the

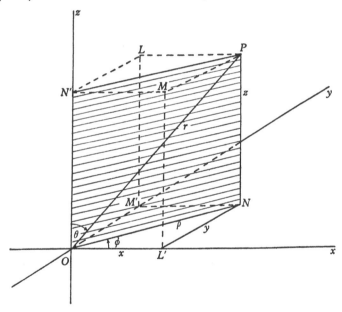

Fig. 104.

'upward vertical'. Through O draw the plane perpendicular to Oz; for reference, we call this plane 'horizontal'.

In the horizontal plane through O take two perpendicular lines Ox, Oy, forming a pair of coordinate axes there in the normal sense of plane Cartesian geometry. (In the diagram, the axis Ox appears to occupy its usual position in the plane—straight across the page.) The three mutually perpendicular lines Ox, Oy, Oz form a set of

Cartesian axes in space, and the position of a point P may be determined by coordinates relative to them, exactly as in the familiar case of two dimensions:

Through P draw lines PL, PM, PN perpendicular to the planes yOz, zOx, xOy respectively. The values of PL, PM, PN, with appropriate signs, fix the position of P precisely, and the values

$$x = \overrightarrow{LP}, \quad y = \overrightarrow{MP}, \quad z = \overrightarrow{NP}$$

are called the CARTESIAN COORDINATES of P for that set of axes.

It is customary to fix the positive senses along the axes by the 'right-handed corkscrew' rule: a corkscrew, turning from positive Oy to positive Oz, is driven along positive Ox; turning from positive Oz to positive Ox, along positive Oy; turning from positive Ox to positive Oy, along positive Oz.

Two other systems in common use may be described by completing the 'box' of which the coordinate planes yOz, zOx, xOy form three faces while P is the vertex opposite to O. The lines PL, PM, PN appear as edges, and the vertices opposite L, M, N are called L', M', N' respectively.

Consider first CYLINDRICAL COORDINATES, in which the point N in the plane xOy, hitherto identified by its Cartesian coordinates x, y, is identified alternatively by its polar coordinates referred to O as pole and Ox as initial line. If we write

$$\rho = \overrightarrow{ON}, \quad \phi = \angle x\overrightarrow{ON}, \quad z = \overrightarrow{NP},$$

then ρ, ϕ, z are called the *cylindrical coordinates* of P.

The angle ϕ may also be thought of as the angle described in the positive sense (as determined by a right-handed corkscrew) when a plane rotates about Oz, from the position zOx into the position zOP.

It is customary to take ρ as positive, though there are times when the unrestricted sign proves more convenient.

Consider next SPHERICAL POLAR COORDINATES, in which the position of P is determined as follows:

(i) it lies in a plane (namely, zOP) making an angle ϕ with a fixed plane (namely, zOx) after rotation about a fixed axis (namely, Oz);

(ii) it lies, in that plane, on a line through the origin O making an angle θ with the fixed axis;

(iii) it is at a distance r from O.

Thus, in the diagram (fig. 104)

$$\phi = \angle x\overrightarrow{ON}, \quad \theta = \angle z\overrightarrow{OP}, \quad r = \overrightarrow{OP}.$$

It is customary to take r as positive, though (as for ρ) the unrestricted sign is sometimes more convenient.

The relations between the sets of coordinates are clear from the diagram. Thus

$$x = \rho \cos \phi, \quad y = \rho \sin \phi;$$

$$x = r \sin \theta \cos \phi, \quad y = r \sin \theta \sin \phi, \quad z = r \cos \theta;$$

$$\rho = \sqrt{(x^2 + y^2)}, \quad r = \sqrt{(x^2 + y^2 + z^2)};$$

$$\phi = \tan^{-1}(y/x), \quad \theta = \tan^{-1}(\rho/z).$$

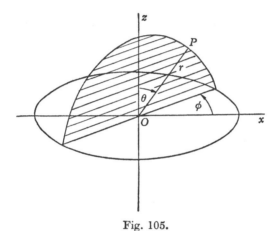

Fig. 105.

An alternative form of diagram (fig. 105) may help to make the 'spherical' nature of the coordinates r, θ, ϕ somewhat clearer. A sphere, of centre O and radius r, cuts the plane xOy in the 'horizontal' circle shown and cuts the plane zOP in a circle of which the 'upper' semicircle is indicated. The 'turning' which defines the angles ϕ, θ may be identified easily.

CHAPTER XIV

PARTIAL DIFFERENTIATION

In the earlier volumes of this book, attention was directed mainly towards the properties of functions of a single variable; we must now examine functions of several. There are, of course, many of these properties which develop by natural extension; but the increased scope afforded by the additional variables introduces difficulties of which the functions readily avail themselves. A searching study, even of quite elementary features, would quickly lead us beyond the limits of the present work, and must be found elsewhere. The intention is to present a picture, accurate as far as it goes, of how the theory extends, so that the reader will be able, on the one hand, to make use of his knowledge, and, on the other, to see later what is involved when he comes to a fuller examination of the details.

We should, however, state at once that it seems necessary in one or two places to relax the standard of rigour at which we aimed for a single independent variable. The language of 'small elements', 'negligible quantities' and so on will be adopted more freely when clarity in the picture might otherwise be sacrificed.

One other point deserves mention. It is now reasonable to assume a greater maturity of outlook and experience than could be expected for the first volumes; in particular, the reader should have some familiarity with more advanced topics such as determinants and coordinate solid geometry. The subject-matter of other branches of mathematics normally studied about the level implied by this volume will usually be incorporated without further reference.

1. The use of geometrical language. Suppose that

$$f(x, y, z, \ldots)$$

is a function of the independent variables

$$x, y, z, \ldots.$$

We shall often picture the numbers $x, y, z, \ldots$ as the coordinates of a point referred to ordinary rectangular Cartesian axes. (Since this will seldom be done except for two variables x, y or three variables

x, y, z, ideas involving space of more than three dimensions will arise only in their most elementary form.) Thus our *analytical* ideas may appear clothed in *geometrical* language.

For example, the function

$$(1/xy),$$

is defined for all non-zero values of x, y, and we may speak of it as 'defined for all points of the (x, y)-plane except on the axes of coordinates'; also, the function

$$\frac{1}{x^2 + y^2 + z^2},$$

is defined for all values of x, y, z not simultaneously zero, and we may speak of it as 'defined for all points of space except the origin'.

Again, we shall meet functions defined only for *limited ranges* of values of the variables. Thus, in a theory of real variables, the function

$$\sqrt{(1 - x^2 - y^2 - z^2)}$$

is 'defined for all points in the region enclosed by the sphere $x^2 + y^2 + z^2 = 1$', and the function

$$\sqrt{(x^2 + y^2 - 4x)}$$

is 'defined for all points outside the circle $x^2 + y^2 = 4x$'. We shall use such language without further explanation when it seems to arise naturally.

2. Continuity. A function $f(x)$ of a single independent variable x is continuous at a point $x = \alpha$ if the difference between $f(x), f(\alpha)$ remains small for *all* values of x sufficiently near to α; and this informal statement of the principle of continuity suggests at once the basis for its extension. For the sake of exposition, consider the case when the number of independent variables is three.

Let

$$f(x, y, z)$$

be a given function of the three independent variables x, y, z, defined for all points in a region of space which includes the point $P(\alpha, \beta, \gamma)$. The function $f(x, y, z)$ is said to be *continuous* at the point (α, β, γ) if, given any positive number ϵ, we can find a positive number η (depending on ϵ) such that the difference

$$| f(x, y, z) - f(\alpha, \beta, \gamma) |$$

remains less than ϵ for all points (x, y, z) whose distance from (α, β, γ) is less than η; that is, if

$$|f(x, y, z) - f(\alpha, \beta, \gamma)| < \epsilon$$

whenever $\quad \sqrt{\{(x - \alpha)^2 + (y - \beta)^2 + (z - \gamma)^2\}} < \eta.$

A slight modification is sometimes useful. We have interpreted 'near (α, β, γ)' as 'lying within a sphere of radius η'; but the sphere may, if we wish, be replaced by a cubical box of edges 2η, giving the alternative inequalities

$$|x - \alpha| < \eta, \quad |y - \beta| < \eta, \quad |z - \gamma| < \eta.$$

As may be expected, the sum, difference and product of two continuous functions can be shown to be continuous; so also is their quotient at any point where the denominator does not vanish.

To illustrate the implications of this concept of continuity, let us consider a function of two variables x, y. The expression

$$\frac{x^2 y^2}{x^2 + y^2}$$

is defined for all values of x, y not both zero, and we take the function $f(x, y)$ based on it, but defined also to have the value zero when $x = y = 0$. Thus

$$f(x, y) = \frac{x^2 y^2}{x^2 + y^2} \quad (x, y \text{ not both zero}),$$

$$= 0 \qquad (x = y = 0).$$

The function is obviously continuous away from the origin, and we now discuss the continuity for $x = y = 0$.

Take a given positive number ϵ, and draw the circle of centre the origin and radius η. Let $Q(x, y)$ be any point, other than the origin, inside the circle. In terms of polar coordinates, we may write

$$x = r \cos \theta, \quad y = r \sin \theta,$$

where $\qquad 0 < r < \eta.$

Then $\qquad f(x, y) = \dfrac{r^4 \sin^2 \theta \, \cos^2 \theta}{r^2}$

$$= r^2 \sin^2 \theta \, \cos^2 \theta,$$

the cancellation being justified since $r > 0$. But

$$r < \eta, \quad \sin^2 \theta \, \cos^2 \theta < 1,$$

and so, for all points (x, y) within the circle,

$$f(x, y) < \eta^2.$$

In particular, if we take $\eta = \sqrt{\epsilon}$, then

$$|f(x, y)| < \epsilon.$$

Hence the function is continuous at the origin, being numerically less than ϵ at *all* points within a distance $\sqrt{\epsilon}$ of it.

As a second example, consider the function defined by the relations

$$f(x, y) = \frac{xy}{x^2 + y^2} \quad (x, y \text{ not both zero})$$

$$= 0 \quad (x = y = 0).$$

To investigate continuity at the origin, take a given positive number ϵ, and draw a circle of centre the origin and radius k. Let $Q(x, y)$ be any point, other than the origin, inside the circle. In terms of polar coordinates we have, as before,

$$f(x, y) = \frac{r^2 \sin \theta \cos \theta}{r^2}$$

$$= \sin \theta \cos \theta.$$

But it is NOT now possible to ensure that this value shall be less than ϵ for all points within the circle; for the function $\sin \theta \cos \theta$, or $\frac{1}{2} \sin 2\theta$, takes all values between $-\frac{1}{2}$ and $\frac{1}{2}$ as θ varies, so that, for given small ϵ, there will certainly be points inside any circle (however small its radius) for which $f(x, y)$ *exceeds* ϵ. For example, $f(x, y) = \frac{1}{2}$ at all points except the origin on the line $y = x$, however close to the origin these points may be. In other words, the number η of the definition cannot be obtained, and the function is discontinuous.

It may be remarked that the formula

$$f(x, y) = \sin \theta \cos \theta$$

shows that the function is, so to speak, 'continuous for approach to the origin along either of the axes', its value for points of either axis being zero; but for approach along any other line there is a discontinuity, namely, a break from the value $\sin \theta \cos \theta$, at points other than the origin, to zero at the origin. Thus it is continuous as a function of x when y is fixed at zero, and vice versa; but not otherwise.

EXAMPLES I

Investigate the continuity at the origin of the following functions, given that, in each case, the value of the function at the origin itself is zero.

1. $\dfrac{x^3 y^3}{x^2 + y^2}$.

2. $\dfrac{1}{x^2 + y^2}$.

3. $\dfrac{x^4 - y^4}{x^4 + y^4}$.

4. $\dfrac{x^2 y^2}{x^4 + y^4}$.

3. Differentiation; introductory example. We have already studied in considerable detail the variation of functions of a single variable. For more than one variable the picture is naturally more complicated, for the values of the function are then affected by changes in each of the variables, and the whole variation arises as a combination of several effects, each making its own contribution to the total.

Some idea of what is involved may be gathered by considering how the volume of a given quantity of gas depends on its temperature and pressure jointly. It is well known that (under suitable conditions) the volume v, the temperature t, and the pressure p are connected by the formula

$$pv = Rt,$$

where R is a numerical constant. Thus

$$v = \frac{Rt}{p}.$$

If t, p undergo small changes to values $t + \delta t$, $p + \delta p$, the volume varies to a value $v + \delta v$, where

$$v + \delta v = \frac{R(t + \delta t)}{p + \delta p},$$

and so, by direct subtraction,

$$\delta v = \frac{R(p\,\delta t - t\,\delta p)}{p(p + \delta p)}.$$

Our concern is, of course, with small variations, and we assume that δt, δp are so small that their squares and their product are negligible in comparison with p. By the binomial theorem we have the *approximate* relation

$$\frac{1}{p+\delta p} = \frac{1}{p}\left(1+\frac{\delta p}{p}\right)^{-1}$$

$$= \frac{1}{p}\left(1-\frac{\delta p}{p}\right),$$

so that
$$\delta v = \frac{R}{p^2}(p\,\delta t - t\,\delta p)\left(1-\frac{\delta p}{p}\right).$$

Neglecting squares and products of small quantities we obtain the formula of variation

$$\delta v = \frac{R}{p^2}(p\,\delta t - t\,\delta p)$$

approximately.

The significance of this formula is that it gives an expression for δv which is *linear* in the increments δt, δp of the independent variables.

There are two special cases which deserve attention:

(i) *Variation at constant pressure.* In an experiment under conditions of constant pressure, the value of the increment δp is actually zero, and so the formula reduces to the relation

$$\delta v = \frac{R}{p}\,\delta t,$$

leading to the relation
$$\frac{\delta v}{\delta t} = \frac{R}{p}.$$

This is, of course, the differential coefficient, with respect to t, of the function $v \equiv Rt/p$, calculated on the assumption that p is constant.

(ii) *Variation at constant temperature.* For constant temperature, the value of δt is zero, so that (approximately)

$$\delta v = -\frac{Rt}{p^2}\,\delta p,$$

or
$$\frac{\delta v}{\delta p} = -\frac{Rt}{p^2},$$

and this is again the differential coefficient, with respect to p, of the function $v \equiv Rt/p$, calculated this time on the assumption that t is constant.

M III

4. Notation. Suppose that

$$w \equiv f(x, y, z)$$

is a function of three independent variables x, y, z. On differentiating w with respect to x, *on the assumption that y, z are treated as constants during the process*, we obtain a differential coefficient which it is customary to write in the form

$$\frac{\partial w}{\partial x}.$$

Similarly the symbols $\dfrac{\partial w}{\partial y}, \dfrac{\partial w}{\partial z}$ denote differentiations with respect to y (on the assumption that z, x are treated as constants) and with respect to z (on the assumption that x, y are treated as constants).

These coefficients are called the PARTIAL DIFFERENTIAL CO-EFFICIENTS *of w with respect to x, y, z respectively.* Other notations in common use are

$$f_x, \quad f_y, \quad f_z$$

and

$$w_x, \quad w_y, \quad w_z.$$

For special purposes, we shall occasionally use the symbols f'_x, f'_y, f'_z.

The partial differential coefficient of $f(x, y, z)$ with respect to x is the limit, if it exists,

$$f_x \equiv \lim_{h \to 0} \frac{f(x+h, y, z) - f(x, y, z)}{h},$$

y, z remaining constant; with similar formulae for f_y, f_z.

ILLUSTRATION 1. If $\quad w = x^4 y^3 z^2,$

then $\quad \dfrac{\partial w}{\partial x} = 4x^3 y^3 z^2, \quad \dfrac{\partial w}{\partial y} = 3x^4 y^2 z^2, \quad \dfrac{\partial w}{\partial z} = 2x^4 y^3 z.$

ILLUSTRATION 2. If $\quad z = e^{ax} \sin by,$

then $\quad \dfrac{\partial z}{\partial x} = a\, e^{ax} \sin by, \quad \dfrac{\partial z}{\partial y} = b\, e^{ax} \cos by.$

EXAMPLES II

Evaluate $\dfrac{\partial z}{\partial x}, \dfrac{\partial z}{\partial y}$ for each of the following functions:

1. $x^2 y^5.$
2. $x^3 / y^2.$
3. $x \sin y.$
4. $\cos xy.$
5. $e^x \cos 2y.$
6. $e^x + e^y.$
7. $x.$
8. $(1+x)^3 e^{-xy}.$
9. $y \sqrt{x}.$

Evaluate $\dfrac{\partial w}{\partial x}, \dfrac{\partial w}{\partial y}, \dfrac{\partial w}{\partial z}$ for each of the following functions:

10. $x^3 y^4/z^2$.   11. $x^4 + y^3 + z^2$.

12. $e^x \sin y \cos z$.   13. $\sin x \sin y \sin z$.

14. $(1+x)^3 e^{yz}$.   15. $x \tan^{-1} yz$.

5. The increment of a function of several variables.

We now apply the ideas illustrated in §3 to functions in general. For conciseness we restrict the statement to functions of two variables, but the extension may easily be completed when required.

Let
$$z \equiv f(x, y)$$
be a given function of two independent variables x, y. Suppose that x, y receive small increments to the values $x + \delta x$, $y + \delta y$ and that z consequently assumes the value $z + \delta z$, where
$$z + \delta z = f(x + \delta x, y + \delta y),$$
so that
$$\delta z = f(x + \delta x, y + \delta y) - f(x, y).$$
It follows, by inserting and cancelling the term $f(x, y + \delta y)$, that
$$\delta z = \{f(x + \delta x, y + \delta y) - f(x, y + \delta y)\} + \{f(x, y + \delta y) - f(x, y)\}.$$

Now the two functions $f(x + \delta x, y + \delta y)$, $f(x, y + \delta y)$ differ only in respect of x, the second variable retaining its value $y + \delta y$ in each. Considered together, they are effectively functions of the single variable x, as if the variable y were reduced temporarily to the status of a parameter. We therefore apply the mean-value theorem for a single variable (Vol. I, p. 61) and obtain the relation
$$f(x + \delta x, y + \delta y) - f(x, y + \delta y) = \delta x f'_x(x + \theta_1 \delta x, y + \delta y),$$
where θ_1 lies between 0, 1, and where the suffix x is inserted at the symbol of differentiation f' to imply that f is differentiated with respect to x only, the second variable meanwhile remaining constant (at value $y + \delta y$).

In the same way, and with similar notation,
$$f(x, y + \delta y) - f(x, y) = \delta y f'_y(x, y + \theta_2 \delta y),$$
where θ_2 lies between 0, 1, and where the suffix y is inserted at the symbol of differentiation f' to imply that f is differentiated with respect to y only, the first variable meanwhile remaining constant (at value x). Thus, in all,
$$\delta z = \delta x f'_x(x + \theta_1 \delta x, y + \delta y) + \delta y f'_y(x, y + \theta_2 \delta y).$$

2-2

[*Note.* For a function w of three variables x, y, z, we should obtain similarly the relation

$$\delta w = \delta x f'_x(x + \theta_1 \delta x, y + \delta y, z + \delta z)$$
$$+ \delta y f'_y(x, y + \theta_2 \delta y, z + \delta z)$$
$$+ \delta z f'_z(x, y, z + \theta_3 \delta z),$$

and so on.]

For example, if $\quad z = f(x,y) \equiv x^3 y^2,$

then $\quad\quad\quad\quad f'_x(x,y) = 3x^2 y^2,$

$$f'_y(x,y) = 2x^3 y,$$

so that $\quad f'_x(x + \theta_1 \delta x, y + \delta y) = 3(x + \theta_1 \delta x)^2 (y + \delta y)^2,$

$$f'_y(x, y + \theta_2 \delta y) = 2x^3(y + \theta_2 \delta y).$$

What we have proved is that, if

$$\delta z \equiv (x + \delta x)^3 (y + \delta y)^2 - x^3 y^2,$$

then there exist numbers θ_1, θ_2, each between 0, 1, such that

$$\delta z = 3(x + \theta_1 \delta x)^2 (y + \delta y)^2 \, \delta x + 2x^3(y + \theta_2 \delta y) \, \delta y.$$

This is as far as we can go with exactitude, but an approximate discussion, when δx, δy are regarded as very small, is of importance for two reasons: first, that it is a case which often occurs in applications; secondly, that the approximation sets the stage for an extension of the idea of *differentials* already used (Vol. I, p. 42) for functions of a single variable.

We now make the assumption that *the two partial differentia coefficients*

$$f'_x(x,y), \quad f'_y(x,y)$$

are continuous functions; we do not want our approximations to be disturbed by 'jumps' in these values. Under this assumption, the difference between the products

$$\delta x f'_x(x + \theta_1 \delta x, y + \delta y), \quad \delta y f'_y(x, y + \theta_2 \delta y)$$

and $\quad\quad\quad\quad \delta x f'_x(x,y), \quad \delta y f'_y(x,y)$

will be very small indeed. We write

$$f'_x(x + \theta_1 \delta x, y + \delta y) - f'_x(x,y) = \epsilon_1,$$
$$f'_y(x, y + \theta_2 \delta y) - f'_y(x,y) = \epsilon_2,$$

where ϵ_1, ϵ_2 both tend to zero as δx, δy tend to zero simultaneously. Thus

$$\delta z = \{f'_x(x,y) + \epsilon_1\}\,\delta x + \{f'_y(x,y) + \epsilon_2\}\,\delta y.$$

The APPROXIMATION then assumes the form

$$\delta z = f'_x(x,y)\,\delta x + f'_y(x,y)\,\delta y,$$

or, in more suggestive notation,

$$\delta z = \frac{\partial z}{\partial x}\,\delta x + \frac{\partial z}{\partial y}\,\delta y.$$

An important feature of this approximation is that the partial differential coefficients $\dfrac{\partial z}{\partial x} \equiv f'_x(x,y)$, $\dfrac{\partial z}{\partial y} \equiv f'_y(x,y)$ are evaluated *at the point* (x,y) *itself.*

6. Geometrical interpretation: Cartesian coordinates.

Let Ox, Oy be the axes for a system of rectangular Cartesian coordinates in a horizontal plane. This is illustrated in the diagram

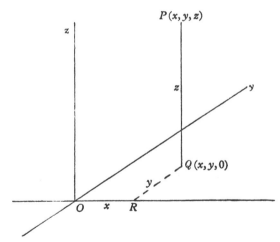

Fig. 106.

(fig. 106), where the reader may regard himself as looking 'down' upon axes drawn in the usual position. Draw a straight line Oz vertically upwards.

Given a point P in space, let the vertical line through it meet the plane xOy in Q, and draw QR perpendicular to Ox. Denote by x, y, z the lengths $\overrightarrow{OR}$, $\overrightarrow{RQ}$, $\overrightarrow{QP}$ respectively; then the triplet x, y, z

may be used as coordinates for the point P in space, just as the pair x, y is used for a point in a plane. If P is the point (x, y, z), then Q is the point $(x, y, 0)$ in the horizontal plane.

The coordinate z gives the height of P referred to xOy as zero level. In particular, if x, y, z are connected by a relation

$$z = f(x, y),$$

where we assume $f(x, y)$ to be a single-valued function defined for each pair of values of x, y, then, as x, y (and consequently z) vary, the point Q moves about the plane xOy, while P describes the

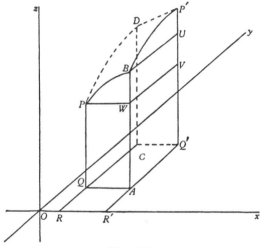

Fig. 107.

surface whose height at any point is equal to the corresponding value of the function. We say that this surface *represents* the function $f(x, y)$.

For instance, it is an easy example on the theorem of Pythagoras to show that the function

$$z = + \sqrt{(1 - x^2 - y^2)}$$

is represented by the hemisphere of centre O and unit radius lying above the plane xOy.

We now assume, for convenience of language, that $\overrightarrow{Ox}$ is due east and $\overrightarrow{Oy}$ due north. We shall also regard the surface $z = f(x, y)$ as a 'hill' and P as the position of a 'climber' moving about on it.

Suppose that the climber is at the point P (fig. 107) defined by the values x, y of the easterly and northerly coordinates, and that he wishes to climb to the point P' defined by $x + \delta x$, $y + \delta y$. The whole crux of the difference between functions of one variable and functions of two lies in the fact that, whereas for one variable the motion along the *curve* which represents the function is defined all the way, for two variables the *surface* may be traversed by an innumerable choice of paths. Moreover, each way of leaving P will demand a gradient all of its own. The partial differential coefficients correspond to selective choices from the available paths.

One way of passing from P to P' which can easily be described in mathematical terms is, first, to move the distance δx easterly to B, and then to move the distance δy northerly. The climber thus describes in succession the two arcs PB, BP' of the diagram.

Let us now suppose that P' is very close to P, so that the arcs PB, BP' are very small. The arc PB may be regarded as almost straight, and so the 'rise' between P and B is proportional to the length δx; say

$$\delta z \,(\text{easterly}) = \alpha \, \delta x.$$

Similarly, BP' is almost straight, and so the 'rise' between B and P' is proportional to δy; say

$$\delta z \,(\text{northerly}) = \beta \, \delta y.$$

If δy is the total 'rise' between P, P', then

$$\delta z = \delta z \,(\text{easterly}) + \delta z \,(\text{northerly})$$
$$= \alpha \, \delta x + \beta \, \delta y.$$

We add that, if the climber had gone first northerly and then easterly, following the course PD, DP' in the diagram, then, for distances so small that the paths may be regarded as straight, $PBP'D$ is (in normal cases) a parallelogram, and so

$$\delta z = \beta \, \delta y + \alpha \, \delta x$$

for the *same* values of α and β.

Two simple observations now complete the illustration. *Geometrically*, α, β are the gradients of those curves which are the sections of the hill in the easterly and northerly directions respectively. *Analytically*, we see, by putting $\delta y = 0$, that α is the ratio $\delta z : \delta x$ calculated on the assumption that y is constant; thus

$$\alpha = \frac{\partial z}{\partial x},$$

and, similarly,
$$\beta = \frac{\partial z}{\partial y}.$$

Hence *the partial differential coefficients* $\dfrac{\partial z}{\partial x}$, $\dfrac{\partial z}{\partial y}$ *are identified as the gradients of the surface* $z = f(x, y)$

in the x- and y-directions respectively.

EXAMPLES III

1. Show that the function

$$z = 1 - \sqrt{(2x - x^2 - y^2)}$$

is represented by an 'inverted' hemisphere.

Prove also that the gradients in the x- and y-directions at the point (x, y, z) are in the ratio $(x - 1) : y$, and that these gradients are equal only for points on a certain diametral plane.

2. Prove that the gradient at the point (x, y, z) of the surface

$$z = 1 - x^2 - 4y^2,$$

for motion in the plane $y = mx$, has the value

$$-2x\,(1 + 4m^2)/\sqrt{(1 + m^2)}.$$

7. Geometrical interpretation: polar coordinates. Referring again to § 6, the position of the point Q in the horizontal plane may be described alternatively by means of polar coordinates r, θ. When this is done, the height z appears as a function of r, θ in the form

$$z = g(r, \theta).$$

The path from P to P' then falls naturally into the two parts PB 'radially' and BP' 'round a circle' (fig. 108). The gradients are quite different from those described in § 6.

If Q, Q' in the plane have coordinates r, θ and $r + \delta r$, $\theta + \delta\theta$, where δr, $\delta\theta$ are regarded as small, the lengths of QA, AQ' do not differ greatly from δr, $r\,\delta\theta$ respectively. Also PB, BP' are almost straight lines.

As before, we indicate the 'rises' by the notation

$$\delta z\,\text{(radial)} \quad \equiv \text{'rise' from } P \text{ to } B,$$

$$\delta z\,\text{(circular)} \equiv \text{'rise' from } B \text{ to } P'.$$

Denoting by α', β' the gradients up PB, BP', we have the relations

$$\delta z\,\text{(radial)} \quad = \alpha' QA \ = \alpha'\,\delta r,$$

$$\delta z\,\text{(circular)} = \beta' AQ' = \beta' r\,\delta\theta,$$

so that the total 'rise' δz is given by the formula

$$\delta z = \alpha' \, \delta r + \beta' r \, \delta \theta.$$

Since $\dfrac{\partial z}{\partial r}$, $\dfrac{\partial z}{\partial \theta}$ are the limiting values of $\delta z \div \delta r$ (θ constant) and $\delta z \div \delta \theta$ (r constant) respectively, we have the formulae

$$\frac{\partial z}{\partial r} = \alpha',$$

$$\frac{\partial z}{\partial \theta} = \beta' r.$$

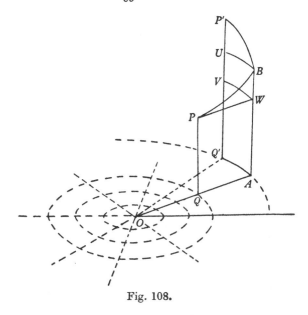

Fig. 108.

Hence the values of the gradients are

$$\frac{\partial z}{\partial r}, \quad \text{radially,}$$

and

$$\frac{1}{r} \frac{\partial z}{\partial \theta}, \quad \text{'circularly'.}$$

ILLUSTRATION 3. Suppose that the height at the point (x, y) is z, where

$$z = ax^2 + by^2.$$

The gradient for motion in the x-direction is

$$\frac{\partial z}{\partial x} \equiv 2ax;$$

and the gradient for motion in the y-direction is

$$\frac{\partial z}{\partial y} \equiv 2by.$$

For 'radial' and 'circular' directions, however, we transform to polar coordinates, so that

$$z = r^2(a \cos^2 \theta + b \sin^2 \theta).$$

The gradient for motion radially is

$$\frac{\partial z}{\partial r} \equiv 2r(a \cos^2 \theta + b \sin^2 \theta)$$

$$\equiv \frac{2}{r}(ax^2 + by^2);$$

and the gradient for motion in the 'circular' direction is

$$\frac{1}{r}\frac{\partial z}{\partial \theta} \equiv r(-2a \cos \theta \sin \theta + 2b \sin \theta \cos \theta)$$

$$\equiv -\frac{2(a-b)xy}{r}.$$

These four gradients are usually quite distinct.

EXAMPLES IV

Evaluate $\dfrac{\partial z}{\partial x}, \dfrac{\partial z}{\partial y}, \dfrac{\partial z}{\partial r}, \dfrac{1}{r}\dfrac{\partial z}{\partial \theta}$ for the surfaces:

1. $z = 2xy.$ 2. $z = x^3 - y^3.$

8. Plane Cartesian and polar coordinates. Suppose that, in a given plane, P is a point which (for convenience) we take in the first quadrant (fig. 109). The Cartesian coordinates x, y and the polar coordinates r, θ are connected by the relations

$$r = \sqrt{(x^2 + y^2)}, \quad \theta = \tan^{-1}(y/x),$$

$$x = r \cos \theta, \quad y = r \sin \theta.$$

Thus r is a function of the two variables x, y, where

$$r = \sqrt{(x^2 + y^2)}.$$

Hence

$$\frac{\partial r}{\partial x} = \frac{x}{\sqrt{(x^2 + y^2)}} = \frac{x}{r},$$

$$\frac{\partial r}{\partial y} = \frac{y}{\sqrt{(x^2 + y^2)}} = \frac{y}{r}.$$

Similarly $\qquad\qquad \theta = \tan^{-1}(y/x),$

so that $\qquad\qquad \dfrac{\partial \theta}{\partial x} = \dfrac{-y}{x^2 + y^2},$

$$\dfrac{\partial \theta}{\partial y} = \dfrac{x}{x^2 + y^2}.$$

The partial differential coefficients of x, y are simpler, namely

$$\dfrac{\partial x}{\partial r} = \cos\theta = \dfrac{x}{r}, \quad \dfrac{\partial y}{\partial r} = \sin\theta = \dfrac{y}{r};$$

and $\qquad \dfrac{\partial x}{\partial \theta} = -r\sin\theta = -y, \quad \dfrac{\partial y}{\partial \theta} = r\cos\theta = x.$

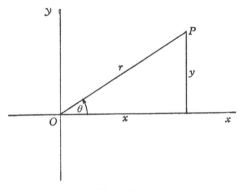

Fig. 109.

[It is important to observe that relations such as $\dfrac{\partial r}{\partial x} = 1 \Big/ \dfrac{\partial x}{\partial r}$ do not hold.]

These partial differentiations may be illustrated graphically. Consider, for example, variation with respect to x, in which y remains constant (fig. 110).

If P moves to a near point P' in such a way that y is constant, the line PP' is parallel to the x-axis. Let the circle with centre O and radius OP cut OP' in U. Then, if δx, δr, $\delta\theta$ are the increments in x, r, θ, we have the relations

$$\delta x = PP', \quad \delta r = UP', \quad \delta\theta = -\angle P'OP,$$

the latter being negative since θ increases in the counter-clockwise sense.

The arc PU cuts the radius OP' at right angles, and, if P' is very close to P, we may regard the arc PU as almost straight. Then PUP' is a right-angled triangle, in which $\angle UP'P = \theta + \delta\theta$. Hence

$$\frac{\delta r}{\delta x} = \cos UP'P = \cos(\theta + \delta\theta)$$

approximately. In the limit, as $\delta x \to 0$, the quotient $\delta r \div \delta x$ tends to the value $\dfrac{\partial r}{\partial x}$, and $\delta\theta$ tends to zero. Hence

$$\frac{\partial r}{\partial x} = \cos\theta = \frac{x}{r},$$

agreeing with the calculated result.

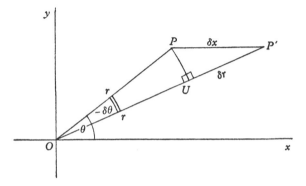

Fig. 110.

Again, the length of the arc UP is $r(-\delta\theta)$, or $-r\,\delta\theta$. Also we have the approximate relation

$$\frac{UP}{PP'} = \sin UP'P,$$

or
$$-\frac{r\,\delta\theta}{\delta x} = \sin(\theta + \delta\theta),$$

or
$$\frac{\delta\theta}{\delta x} = -\frac{1}{r}\sin(\theta + \delta\theta).$$

In the limit, as $\delta x \to 0$, we have the relation

$$\frac{\partial\theta}{\partial x} = -\frac{1}{r}\sin\theta$$

$$= -\frac{y}{r^2},$$

again agreeing with the calculated result.

To illustrate partial differentiation with respect to r, let us consider the coefficient $\dfrac{\partial x}{\partial r}$ (fig. 111). This requires a variation from P to P' in which θ remains constant, so that P' lies along the radius OP. It is then an easy matter in elementary trigonometry to derive, for increments δx, δr, the relation

$$\frac{\delta x}{\delta r} = \cos \theta,$$

so that, as $\delta r \to 0$, $\dfrac{\partial x}{\partial r} = \cos \theta = \dfrac{x}{r}.$

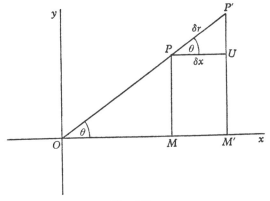

Fig. 111.

Finally, consider, for partial differentiation with respect to θ, the coefficient $\dfrac{\partial x}{\partial \theta}$. This requires a variation from P to P' in which r remains constant, so that P' lies on the circle of centre O and radius OP (fig. 112). Draw PM, $P'M'$ perpendicular to the axis OX, and PU perpendicular to $P'M'$. Then the increments $\delta\theta$, δx are given by the relations $\delta\theta = \angle POP',$

$$-\delta x = M'M,$$

the minus sign arising since δx is negative. The length of PP' is $r\,\delta\theta$, and the radius OP is at right angles to the arc PP'.

For the interpretation, we suppose that P' is very near to P, so that PP' is almost a straight line. Then

$$\angle UPP' = \tfrac{1}{2}\pi - \angle UPO = \tfrac{1}{2}\pi - \theta.$$

But
$$\frac{UP}{PP'} = \cos UPP',$$

so that
$$\frac{-\delta x}{r\,\delta\theta} = \sin\theta$$

approximately. Hence in the limit, as $\delta\theta \to 0$,

$$\frac{\partial x}{\partial \theta} = -r\sin\theta,$$

as calculated.

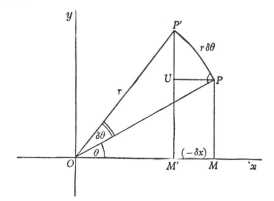

Fig. 112.

EXAMPLES V

Illustrate graphically the relations

1. $\dfrac{\partial r}{\partial y} = \sin\theta.$ 2. $\dfrac{\partial \theta}{\partial y} = \dfrac{x}{r^2}.$

3. $\dfrac{\partial y}{\partial r} = \dfrac{y}{r}.$ 4. $\dfrac{\partial y}{\partial \theta} = r\cos\theta.$

9. Connected 'independent' variables. It often happens that a function
$$z \equiv f(x, y)$$

is given in terms of two variables x, y, but that those variables are in fact related, each being a function of a single variable t (which might, in a special case, be x or y). Then z is a function of t, say

$$z \equiv F(t).$$

[For example, the point (x, y) might be restricted to lie on a curve, such as the parabola $x = at^2$, $y = 2at$.]

To find an expression for the differential coefficient

$$\frac{dz}{dt} \equiv F'(t),$$

assuming the existence and continuity of the coefficients $\frac{\partial f}{\partial x}, \frac{\partial f}{\partial y}$, *and the existence of* $\frac{dx}{dt}, \frac{dy}{dt}$.

Suppose that t receives an increment u, as a result of which x, y become $x+h$, $y+k$. Then

$$\frac{F(t+u)-F(t)}{u} = \frac{f(x+h,y+k)-f(x,y)}{u}$$

$$= \frac{\{f(x+h,y+k)-f(x,y+k)\}+\{f(x,y+k)-f(x,y)\}}{u}$$

$$= \frac{hf'_x(x+\theta_1 h,y+k)+kf'_y(x,y+\theta_2 k)}{u},$$

as for the similar argument on p. 11. Hence

$$\frac{F(t+u)-F(t)}{u} = \frac{h}{u}f'_x(x+\theta_1 h,y+k)+\frac{k}{u}f'_y(x,y+\theta_2 k).$$

Now proceed to the limit, letting u tend to zero. Since $\frac{dx}{dt}, \frac{dy}{dt}$ exist,

$$\lim_{u\to 0}\frac{h}{u} = \frac{dx}{dt}, \quad \lim_{u\to 0}\frac{k}{u} = \frac{dy}{dt};$$

also h, k both tend to zero with u. Hence, by the usual theorems on sums and products of limits (Vol. I, p. 25) and the continuity of $\frac{\partial f}{\partial x}, \frac{\partial f}{\partial y}$, the relation becomes

$$\frac{dF}{dt} = \frac{dx}{dt}f'_x(x,y)+\frac{dy}{dt}f'_y(x,y),$$

or, on rewriting,
$$\frac{dF}{dt} = \frac{\partial f}{\partial x}\frac{dx}{dt} + \frac{\partial f}{\partial y}\frac{dy}{dt}.$$

This relation is also written in the form

$$\frac{dz}{dt} = \frac{\partial z}{\partial x}\frac{dx}{dt} + \frac{\partial z}{\partial y}\frac{dy}{dt},$$

where it is understood that z is expressed in terms of t on the left and of x, y on the right.

ILLUSTRATION 4. Suppose that z is the square of the distance of a point $P(x, y)$ on the ellipse

$$\frac{x^2}{a^2} + \frac{y^2}{b^2} = 1$$

from the focus $(ae, 0)$. Then

$$z = (x - ae)^2 + y^2.$$

The variables x, y are not independent, but can be expressed in terms of a parameter t in the form

$$x = a \cos t, \quad y = b \sin t.$$

Then
$$\frac{\partial z}{\partial x} = 2(x - ae) = 2a(\cos t - e),$$

$$\frac{\partial z}{\partial y} = 2y = 2b \sin t,$$

$$\frac{dx}{dt} = -a \sin t, \quad \frac{dy}{dt} = b \cos t.$$

Hence, by the formula,

$$\frac{dz}{dt} = -2a^2(\cos t - e) \sin t + 2b^2 \sin t \cos t$$
$$= 2a^2 e \sin t - 2(a^2 - b^2) \sin t \cos t$$
$$= 2a^2 e \sin t - 2a^2 e^2 \sin t \cos t.$$

For example, the value of z is a maximum or minimum when $\frac{dz}{dt} = 0$; that is, when
$$\sin t = 0,$$
or
$$\cos t = 1/e.$$

The latter solution is impossible (since $e < 1$), and so the only turning values occur when $\sin t = 0$; that is, when P is at an end of the major axis.

ILLUSTRATION 5. Suppose that

$$z \equiv f(x, y),$$

where x, y are *linear* functions of the variable t; say,

$$x = a + pt, \quad y = b + qt.$$

Then
$$\frac{dz}{dt} = \frac{\partial z}{\partial x}\frac{dx}{dt} + \frac{\partial z}{\partial y}\frac{dy}{dt}$$

$$= p\frac{\partial z}{\partial x} + q\frac{\partial z}{\partial y}.$$

EXAMPLES VI

Evaluate $F'(t)$ by the rule given in this section, and verify your answer by calculating $F(t)$ explicitly and then differentiating it, for each of the following examples:

1. $f(x,y) = x^4 y^2;$ $x = t^2,$ $y = t^4.$

2. $f(x,y) = x\,e^y;$ $x = \cos t,$ $y = \sin t.$

3. $f(x,y) = x^2 + y^2;$ $x = \cos t,$ $y = \sin t.$

4. $f(x,y) = x^2 e^{xy};$ $x = t^2,$ $y = 2t.$

10. The chain rule. The work of the preceding paragraph may be generalized. Suppose that the two* variables x, y are each functions of the three* variables ξ, η, ζ; say $x(\xi, \eta, \zeta), y(\xi, \eta, \zeta)$. Then a given function $w(x,y)$ of the variables x, y can be expressed in terms of ξ, η, ζ as a function $\rho(\xi, \eta, \zeta)$. We obtain *an expression for the partial differential coefficient*

$$\frac{\partial \rho}{\partial \xi},$$

in terms of the partial differential coefficients of w, x, y.

When ξ, η, ζ receive increments $\delta\xi$, $\delta\eta$, $\delta\zeta$, the variables x, y become $x + \delta x$, $y + \delta y$ respectively, and $w(x,y)$ becomes

$$w(x + \delta x, y + \delta y).$$

The increment in w (or ρ) is (compare p. 11)

$$w(x + \delta x, y + \delta y) - w(x,y)$$
$$= \delta x\, w'_x(x + \theta_1 \delta x, y + \delta y) + \delta y\, w'_y(x, y + \theta_2 \delta y). \quad (1)$$

Moreover, x, y are themselves functions of ξ, η, ζ, so that the corresponding increments are

$$\delta x \equiv x(\xi + \delta\xi, \eta + \delta\eta, \zeta + \delta\zeta) - x(\xi, \eta, \zeta)$$
$$= \delta\xi\, x'_\xi(\xi + \phi_1 \delta\xi, \eta + \delta\eta, \zeta + \delta\zeta)$$
$$+ \delta\eta\, x'_\eta(\xi, \eta + \phi_2 \delta\eta, \zeta + \delta\zeta)$$
$$+ \delta\zeta\, x'_\zeta(\xi, \eta, \zeta + \phi_3 \delta\zeta), \quad (2)$$

with a similar expression in y.

---

* The numbers, two, three, are only chosen for illustration and may be any positive integers.

         M III

On substituting from equation (2), and the corresponding equation with y, into equation (1), we obtain as the coefficient of $\delta\xi$ in the increment of w the expression

$$w'_x(x+\theta_1\delta x, y+\delta y)\, x'_\xi(\xi+\phi_1\delta\xi, \eta+\delta\eta, \zeta+\delta\zeta)$$
$$+ w'_y(x, y+\theta_2\delta y)\, y'_\xi(\xi+\psi_1\delta\xi, \eta+\delta\eta, \zeta+\delta\zeta),$$

where the number ψ_1 replaces the ϕ_1 of equation (2).

Now restrict our attention to those cases in which the partial differential coefficients are continuous. Then (p. 13)

$$w'_x(x+\theta_1\delta x, y+\delta y) = w'_x(x,y)+\epsilon_1 \quad = \frac{\partial w}{\partial x}+\epsilon_1,$$

$$w'_y(x, y+\theta_2\delta y) = w'_y(x,y)+\epsilon_2 \quad = \frac{\partial w}{\partial y}+\epsilon_2,$$

$$x'_\xi(\xi+\phi_1\delta\xi, \eta+\delta\eta, \zeta+\delta\zeta) = x'_\xi(\xi,\eta,\zeta)+\lambda_1 = \frac{\partial x}{\partial\xi}+\lambda_1,$$

$$y'_\xi(\xi+\psi_1\delta\xi, \eta+\delta\eta, \zeta+\delta\zeta) = y'_\xi(\xi,\eta,\zeta)+\lambda_2 = \frac{\partial y}{\partial\xi}+\lambda_2,$$

where λ_1, λ_2 tend to zero as $\delta\xi, \delta\eta, \delta\zeta$ tend to zero; so that $\delta x, \delta y$ also tend to zero, and hence ϵ_1, ϵ_2 also. Thus the increment in w (or, to be more exact, in ρ, which is the same as w but with ξ, η, ζ as the independent variables) is $\delta\rho$, where

$$\delta\rho = \left(\frac{\partial w}{\partial x}\frac{\partial x}{\partial\xi}+\frac{\partial w}{\partial y}\frac{\partial y}{\partial\xi}\right)\delta\xi$$

$$+ \left(\frac{\partial w}{\partial x}\frac{\partial x}{\partial\eta}+\frac{\partial w}{\partial y}\frac{\partial y}{\partial\eta}\right)\delta\eta$$

$$+ \left(\frac{\partial w}{\partial x}\frac{\partial x}{\partial\zeta}+\frac{\partial w}{\partial y}\frac{\partial y}{\partial\zeta}\right)\delta\zeta,$$

plus terms which vanish with $\delta\xi, \delta\eta, \delta\zeta$.

Hence we have the relations, forming the *chain rule,*

$$\frac{\partial\rho}{\partial\xi} = \frac{\partial w}{\partial x}\frac{\partial x}{\partial\xi}+\frac{\partial w}{\partial y}\frac{\partial y}{\partial\xi},$$

$$\frac{\partial\rho}{\partial\eta} = \frac{\partial w}{\partial x}\frac{\partial x}{\partial\eta}+\frac{\partial w}{\partial y}\frac{\partial y}{\partial\eta},$$

$$\frac{\partial\rho}{\partial\zeta} = \frac{\partial w}{\partial x}\frac{\partial x}{\partial\zeta}+\frac{\partial w}{\partial y}\frac{\partial y}{\partial\zeta}.$$

More generally:

If w is a function of the m variables $x_1, x_2, \ldots, x_m$ which are themselves functions of the n variables $\xi_1, \xi_2, \ldots, \xi_n$, and if w when expressed in terms of $\xi_1, \xi_2, \ldots, \xi_n$ is a function ρ, then

$$\frac{\partial \rho}{\partial \xi_i} = \sum_{j=1}^{m} \frac{\partial \rho}{\partial x_j} \frac{\partial x_j}{\partial \xi_i} \quad (i = 1, 2, \ldots, n).$$

ILLUSTRATION 6. Let $\quad w \equiv xy^2z^3,$

where $\quad x = \cos\theta \sin\phi, \quad y = \sin\theta \sin\phi, \quad z = \cos\phi.$

Then

$$\frac{\partial w}{\partial \theta} = \frac{\partial w}{\partial x}\frac{\partial x}{\partial \theta} + \frac{\partial w}{\partial y}\frac{\partial y}{\partial \theta} + \frac{\partial w}{\partial z}\frac{\partial z}{\partial \theta}$$

$$= -y^2z^3 \sin\theta \sin\phi + 2xyz^3 \cos\theta \sin\phi + 3xy^2z^2 . 0$$

$$= -\sin^3\theta \sin^3\phi \cos^3\phi + 2\cos^2\theta \sin\theta \sin^3\phi \cos^3\phi$$

$$= (2 - 3\sin^2\theta)\sin\theta \sin^3\phi \cos^3\phi;$$

$$\frac{\partial w}{\partial \phi} = \frac{\partial w}{\partial x}\frac{\partial x}{\partial \phi} + \frac{\partial w}{\partial y}\frac{\partial y}{\partial \phi} + \frac{\partial w}{\partial z}\frac{\partial z}{\partial \phi}$$

$$= y^2z^3 \cos\theta \cos\phi + 2xyz^3 \sin\theta \cos\phi - 3xy^2z^2 \sin\phi$$

$$= \sin^2\theta \cos\theta \sin^2\phi \cos^4\phi + 2\sin^2\theta \cos\theta \sin^2\phi \cos^4\phi$$
$$\quad - 3\sin^2\theta \cos\theta \sin^4\phi \cos^2\phi$$

$$= 3\sin^2\theta \cos\theta \sin^2\phi \cos^2\phi(\cos^2\phi - \sin^2\phi).$$

EXAMPLES VII

1. Given that $\qquad w = xyz,$

where $\qquad x = \cos\theta \sin\phi, \quad y = \sin\theta \sin\phi, \quad z = \cos\phi,$

evaluate $\qquad \dfrac{\partial w}{\partial \theta}, \quad \dfrac{\partial w}{\partial \phi}.$

2. Given that $\qquad w = u^2 - v^2,$

where $\qquad u = x + y + z, \quad v = xyz,$

evaluate $\qquad \dfrac{\partial w}{\partial x}, \quad \dfrac{\partial w}{\partial y}, \quad \dfrac{\partial w}{\partial z}.$

3. Given that $\qquad w = e^u \cos v,$

where $\qquad u = x + y, \quad v = x - y,$

evaluate $\qquad \dfrac{\partial w}{\partial x}, \quad \dfrac{\partial w}{\partial y}.$

4. Given that $\qquad w = x^3 \log yz,$

where $\qquad x = u^2 + v^2, \quad y = u+v, \quad z = u-v,$

evaluate $\qquad \dfrac{\partial w}{\partial u}, \quad \dfrac{\partial w}{\partial v}.$

11. Differentials. We now transfer our attention from the approximate theory of small increments to a more exact theory based on the concept of differentials.

Suppose that $\qquad z \equiv f(x,y)$

is a function of the two independent variables x, y, and that these variables receive increments δx, δy *not necessarily small*. Since we know what is meant by the partial differential coefficients $\dfrac{\partial z}{\partial x}$, $\dfrac{\partial z}{\partial y}$ (see § 4), we may form the function

$$\frac{\partial z}{\partial x}\delta x + \frac{\partial z}{\partial y}\delta y,$$

which, for given increments, has a definite value. It is called *the* DIFFERENTIAL *of z for the increments* δx, δy, and is denoted by the symbol dz, so that, as a matter of definition,

$$dz = \frac{\partial z}{\partial x}\delta x + \frac{\partial z}{\partial y}\delta y.$$

In particular, we require meanings for the differentials of x, y themselves. If we regard x as a function of the two independent variables x, y, then

$$\frac{\partial x}{\partial x} = 1, \quad \frac{\partial x}{\partial y} = 0,$$

so that $\qquad dx = 1.\delta x + 0.\delta y$

$$= \delta x.$$

Similarly $\qquad dy = \delta y.$

Thus the differentials of the two functions x, y for the increments δx, δy are just δx, δy respectively. Hence for given increments δx, δy the differentials dx, dy, dz are connected by the (exact) relation

$$dz = \frac{\partial z}{\partial x}dx + \frac{\partial z}{\partial y}dy.$$

Returning for a moment to the geometrical illustration of § 6 (pp. 13–16), we can give an intuitive interpretation of the differentials, though a precise statement would be hard at present. The surface (fig. 107) $z = f(x, y)$ meets the vertical lines $BA, P'Q'$ at the points B, P'; suppose that the 'tangent plane' at P meets them in $\bar{B}, \bar{P}'$. The sections of the tangent plane in the easterly and northerly directions really *are* straight lines, and their gradients are exactly

$$\frac{\partial z}{\partial x}, \quad \frac{\partial z}{\partial y}$$

calculated at P. Hence the rise from P to $\bar{P}'$ is precisely

$$\frac{\partial z}{\partial x} \delta x + \frac{\partial z}{\partial y} \delta y,$$

or, by definition, the differential dz. Thus *the differential dz is the increment in z when the surface $z = f(x, y)$ is replaced by the tangent plane to it at P*. For small increments, the two values are approximately equal.

12. The differential of a function of two DEPENDENT variables, in the form $dz = \dfrac{\partial z}{\partial x} dx + \dfrac{\partial z}{\partial y} dy$.

With the notation of § 9 (p. 22), multiply the relation proved there, namely

$$\frac{dz}{dt} = \frac{\partial z}{\partial x} \frac{dx}{dt} + \frac{\partial z}{\partial y} \frac{dy}{dt},$$

by the differential dt. Then

$$\frac{dz}{dt} dt = \frac{\partial z}{\partial x} \frac{dx}{dt} dt + \frac{\partial z}{\partial y} \frac{dy}{dt} dt.$$

Now z (on the left), x, y are functions of the single variable t, and we know (Vol. I, p. 43) that their differentials dz, dx, dy are defined* by the relations

$$dz = \frac{dz}{dt} dt, \quad dx = \frac{dx}{dt} dt, \quad dy = \frac{dy}{dt} dt,$$

where $\dfrac{dx}{dt}, \dfrac{dy}{dt}, \dfrac{dz}{dt}$ are differential coefficients with respect to t. Hence the relation assumes the simple form

$$dz = \frac{\partial z}{\partial x} dx + \frac{\partial z}{\partial y} dy,$$

giving the connexion between the differentials dx, dy, dz.

* It should be clearly understood that we are appealing directly to the definitions. We are *not* merely cancelling dt, though the notation dz/dt for a differential coefficient may create an optical illusion to that effect.

The significance of this result will be appreciated on referring to § 11 (p. 28), where precisely this formula appears under the hypothesis that the variables x, y are independent. The work of the two paragraphs thus unites to obtain the important theorem that, *if z is a function of the two variables x, y, then the relation*

$$dz = \frac{\partial z}{\partial x}\,dx + \frac{\partial z}{\partial y}\,dy$$

is true both when the variables x, y on the right are independent, and when they are dependent.

13. Digression on implicit functions. A detailed study of the properties of implicit functions is difficult, and cannot be attempted here. The present paragraph seeks to illustrate, mainly by example, the ideas which the reader should have at the back of his mind in the work which follows.

Consider the relation $xy^3z^5 - 1 = 0$.

It can be used in three ways:

   (i) to express x as a function of y, z in the form

$$x = y^{-3}z^{-5};$$

   (ii) to express y as a function of z, x in the form

$$y = z^{-\frac{5}{3}}x^{-\frac{1}{3}};$$

   (iii) to express z as a function of x, y in the form

$$z = x^{-\frac{1}{5}}y^{-\frac{3}{5}}.$$

These three functional relations are *implicit* in the given equation, and we have, in fact, been able to make them *explicit* by direct solution.

There are, of course, many implicit relations for which direct solution is (in practice, if not in theory) quite impossible; for example $e^{xyz} - \sin(x+y+z) = 0$.

In any event, there are often advantages in keeping the equation in its implicit form, without direct solution, and so the corresponding rules for partial differentiation must be formulated.

Returning to the example $xy^3z^5 = 1$, we obtain the differential relations corresponding to the solutions (i), (ii), (iii) in the forms

$$dx = -3y^{-4}z^{-5}\,dy - 5y^{-3}z^{-6}\,dz,$$
$$dy = -\tfrac{5}{3}z^{-\frac{8}{3}}x^{-\frac{1}{3}}\,dz - \tfrac{1}{3}z^{-\frac{5}{3}}x^{-\frac{4}{3}}\,dx,$$
$$dz = -\tfrac{1}{5}x^{-\frac{6}{5}}y^{-\frac{3}{5}}\,dx - \tfrac{3}{5}x^{-\frac{1}{5}}y^{-\frac{8}{5}}\,dy,$$

respectively.

Multiply the first equation on the left by x^{-1} and on the right by its equivalent y^3z^5; then

$$x^{-1}dx = -3y^{-1}dy - 5z^{-1}dz.$$

Multiply the second equation on the left by $3y^{-1}$ and on the right by its equivalent $3z^{\frac{3}{5}}x^{\frac{1}{3}}$; then

$$3y^{-1}dy = -5z^{-1}dz - x^{-1}dx.$$

Multiply the third equation on the left by $5z^{-1}$ and on the right by its equivalent $x^{\frac{1}{5}}y^{\frac{3}{5}}$; then

$$5z^{-1}dz = -x^{-1}dx - 3y^{-1}dy.$$

It is immediately obvious that the three differential relations just obtained are identical. Indeed, they could have been obtained at once by writing down the differential du of the function

$$u = xy^3z^5 - 1$$

in the form $du = y^3z^5\,dx + 3xy^2z^5\,dy + 5xy^3z^4\,dz,$

and noting that $du = 0$ since u has the constant value zero.

This example illustrates a technique which we shall adopt without further comment. Suppose that, say, three variables x, y, z are connected implicitly by a functional relation

$$u(x, y, z) = 0.$$

Then we obtain the same relation between the differentials dx, dy, dz whether we find it directly by equating du to zero or by first expressing x as a function of y, z (or y of z, x; or z of x, y).

14. Differentials of functionally related variables. Suppose that three variables x, y, z are connected by some relation, so that either may be expressed as a function of the two others. Regarding x as a function of y, z, we have

$$dx = \frac{\partial x}{\partial y}dy + \frac{\partial x}{\partial z}dz;$$

regarding y as a function of z, x, we have

$$dy = \frac{\partial y}{\partial z}dz + \frac{\partial y}{\partial x}dx;$$

regarding z as a function of x, y, we have

$$dz = \frac{\partial z}{\partial x}dx + \frac{\partial z}{\partial y}dy.$$

(As we have just explained, *these three forms of the relation are in fact identical*, the ratios $dx : dy : dz$ being the same in each.)

An alternative notation for the partial differential coefficients is sometimes found useful as giving greater precision. We write, for example,

$$\frac{\partial x}{\partial y}\bigg)_z$$

to denote the result of differentiating x partially with respect to y when the other independent variable is z. With such notation, the three relations given above are

$$dx = \frac{\partial x}{\partial y}\bigg)_z dy + \frac{\partial x}{\partial z}\bigg)_y dz,$$

$$dy = \frac{\partial y}{\partial z}\bigg)_x dz + \frac{\partial y}{\partial x}\bigg)_z dx,$$

$$dz = \frac{\partial z}{\partial x}\bigg)_y dx + \frac{\partial z}{\partial y}\bigg)_x dy.$$

[Thus if the implicit relation is

$$x^2 + y^2 + z^2 = 1,$$

then

$$\frac{\partial x}{\partial y}\bigg)_z = -\frac{y}{x}, \quad \frac{\partial x}{\partial z}\bigg)_y = -\frac{z}{x},$$

$$\frac{\partial y}{\partial z}\bigg)_x = -\frac{z}{y}, \quad \frac{\partial y}{\partial x}\bigg)_z = -\frac{x}{y},$$

$$\frac{\partial z}{\partial x}\bigg)_y = -\frac{x}{z}, \quad \frac{\partial z}{\partial y}\bigg)_x = -\frac{y}{z}.]$$

The three ways of looking at the functional relation, and consequent three ways of looking at the differential relation, all result in a *linear* equation among the differentials dx, dy, dz; and it is often convenient to express this linear relation with symmetric notation

$$u\,dx + v\,dy + w\,dz = 0,$$

(where the symbols u, v, w denote certain functions of the variables x, y, z) leaving unspecified at this stage which of the variables are to be regarded as independent and which as dependent.

With this notation, the partial differential coefficients are

$$\frac{\partial x}{\partial y}\bigg)_z = -\frac{v}{u}, \quad \frac{\partial x}{\partial z}\bigg)_y = -\frac{w}{u},$$

$$\frac{\partial y}{\partial z}\bigg)_x = -\frac{w}{v}, \quad \frac{\partial y}{\partial x}\bigg)_z = -\frac{u}{v},$$

$$\frac{\partial z}{\partial x}\bigg)_y = -\frac{u}{w}, \quad \frac{\partial z}{\partial y}\bigg)_x = -\frac{v}{w}.$$

This leads to another observation. Consider, for example, the coefficient
$$\frac{\partial x}{\partial y}\bigg)_z.$$

It is calculated by differentiation on the assumption that z (the other independent variable) is constant. and this assumption may be stated in the equivalent form that dz is zero. But then the differential relation $u\,dx + v\,dy + w\,dz = 0$ becomes simply
$$u\,dx + v\,dy = 0,$$

and it follows, since
$$\frac{\partial x}{\partial y}\bigg)_z = -\frac{v}{u},$$

that
$$\frac{\partial x}{\partial y}\bigg)_z = dx \div dy.$$

In other words, *the partial differential coefficient $\dfrac{\partial x}{\partial y}\bigg)_z$, where z is kept constant, is equal to the ratio $dx \div dy$ of the differentials dx, dy when dz is equated to zero.*

These ideas may be extended to any number of variables, and the resulting theorems are important in practice:

(i) If z is a function of the variables $x_1, x_2, ..., x_n$, then
$$dz = \frac{\partial z}{\partial x_1}\,dx_1 + \frac{\partial z}{\partial x_2}\,dx_2 + ... + \frac{\partial z}{\partial x_n}\,dx_n$$
$$= \sum_1^n \frac{\partial z}{\partial x_i}\,dx_i,$$

whether the variables $x_1, x_2, ..., x_n$ are independent or not.

(ii) If there is a functional relation between n variables $y_1, y_2, ..., y_n$, then their differentials are correspondingly connected by a linear relation of the form
$$u_1\,dy_1 + u_2\,dy_2 + ... + u_n\,dy_n = 0,$$

or
$$\sum_1^n u_i\,dy_i = 0,$$

where $u_1, u_2, ..., u_n$ are functions of the n variables.

Further, if we regard, say, y_1 as a function of the remaining variables $y_2, y_3, ..., y_n$, then the partial differential coefficient
$$\frac{\partial y_1}{\partial y_2}$$

(calculated with $y_3, y_4, ..., y_n$ all constant) is equal to the ratio
$$dy_1 \div dy_2$$

calculated with $dy_3 = dy_4 = \ldots = dy_n = 0$; so that

$$\frac{\partial y_1}{\partial y_2} = -\frac{u_2}{u_1}.$$

ILLUSTRATION 7. *Three variables x, y, z are such that each may be regarded as a function of the other two. To prove that*

$$\left(\frac{\partial x}{\partial y}\right)_z \left(\frac{\partial y}{\partial x}\right)_z = 1,$$

and that

$$\left(\frac{\partial y}{\partial z}\right)_x \left(\frac{\partial z}{\partial x}\right)_y \left(\frac{\partial x}{\partial y}\right)_z = -1,$$

where, for example, $\left(\dfrac{\partial x}{\partial y}\right)_z$ *means that x is expressed as a function of y, z, and then z kept constant in the differentiation.*

A relation between x, y, z implies a linear relation between their differentials. Thus functions a, b, c of the variables exist such that

$$a\,dx + b\,dy + c\,dz = 0.$$

Hence
$$\left(\frac{\partial x}{\partial y}\right)_z = dx \div dy \quad \text{when} \quad dz = 0$$
$$= -b/a,$$

and
$$\left(\frac{\partial y}{\partial x}\right)_z = dy \div dx \quad \text{when} \quad dz = 0$$
$$= -a/b,$$

so that
$$\left(\frac{\partial x}{\partial y}\right)_z \left(\frac{\partial y}{\partial x}\right)_z = 1.$$

In the same way,
$$\left(\frac{\partial y}{\partial z}\right)_x = -c/b,$$

$$\left(\frac{\partial z}{\partial x}\right)_y = -a/c,$$

$$\left(\frac{\partial x}{\partial y}\right)_z = -b/a,$$

so that
$$\left(\frac{\partial y}{\partial z}\right)_x \left(\frac{\partial z}{\partial x}\right)_y \left(\frac{\partial x}{\partial y}\right)_z = -1.$$

ILLUSTRATION 8. *Four variables u, t, p, v have the property that each one of them can be expressed as a function of any two of the others. To prove that*

$$\left(\frac{\partial u}{\partial t}\right)_p = \left(\frac{\partial u}{\partial t}\right)_v + \left(\frac{\partial u}{\partial v}\right)_t \left(\frac{\partial v}{\partial t}\right)_p,$$

where, for example, $\left(\dfrac{\partial u}{\partial t}\right)_p$ *means that u is expressed as a function of t, p, and then p kept constant in the differentiation.*

A relation between three variables implies a linear relation between their differentials. Since u, t, p are related, functions a, b, c of the variables exist such that

$$a\,du + b\,dt + c\,dp = 0.$$

Thus
$$\left(\frac{\partial u}{\partial t}\right)_p = du \div dt \quad \text{when} \quad dp = 0$$
$$= -b/a.$$

There is also a relation between u, t, v, and functions A, B, C of the variables therefore exist such that

$$A\,du + B\,dt + C\,dv = 0.$$

Thus
$$\left(\frac{\partial u}{\partial t}\right)_v = du \div dt \quad \text{when} \quad dv = 0$$
$$= -B/A,$$

and
$$\left(\frac{\partial u}{\partial v}\right)_t = du \div dv \quad \text{when} \quad dt = 0$$
$$= -C/A.$$

This accounts for the first three partial differential coefficients in the relation which we have to prove. There remains $\left(\dfrac{\partial v}{\partial t}\right)_p$, for which we need the relation between dv, dt, dp. This is found by eliminating du between the two differential relations, giving

$$\frac{b\,dt + c\,dp}{a} = -du = \frac{B\,dt + C\,dv}{A},$$

or
$$aC\,dv + (aB - bA)\,dt - cA\,dp = 0.$$

Hence
$$\left(\frac{\partial v}{\partial t}\right)_p = dv \div dt \quad \text{when} \quad dp = 0$$
$$= \frac{bA - aB}{aC} = \frac{b}{a}\frac{A}{C} - \frac{B}{C}$$
$$= \frac{\left(\dfrac{\partial u}{\partial t}\right)_p}{\left(\dfrac{\partial u}{\partial v}\right)_t} - \frac{\left(\dfrac{\partial u}{\partial t}\right)_v}{\left(\dfrac{\partial u}{\partial v}\right)_t},$$

so that
$$\left(\frac{\partial u}{\partial t}\right)_p = \left(\frac{\partial u}{\partial t}\right)_v + \left(\frac{\partial u}{\partial v}\right)_t \left(\frac{\partial v}{\partial t}\right)_p.$$

EXAMPLES VIII

1. Prove that, if u, y are each functions of the two variables x, t, and if u is expressed (by elimination of t) as a function of x, y, then

$$\left(\frac{\partial u}{\partial y}\right)_x = \left(\frac{\partial u}{\partial t}\right)_x \bigg/ \left(\frac{\partial y}{\partial t}\right)_x.$$

Illustrate this result by considering the cases

(i) $u = x^2 + t^2$, $y = xt$;

(ii) $u = e^x \cos t$, $y = e^x \sin t$.

2. Each of the variables u, v, w is a function of the two variables x, y. By elimination of x, y, the variable u is expressed as a function of v, w. Prove that

$$\left(\frac{\partial u}{\partial w}\right)_v = \frac{\dfrac{\partial u}{\partial x}\dfrac{\partial v}{\partial y} - \dfrac{\partial u}{\partial y}\dfrac{\partial v}{\partial x}}{\dfrac{\partial w}{\partial x}\dfrac{\partial v}{\partial y} - \dfrac{\partial w}{\partial y}\dfrac{\partial v}{\partial x}}.$$

Illustrate this result by considering the cases

(i) $u = x^2 + y^2$, $v = x^2 - y^2$, $w = 2xy$.

(ii) $u = x^3 + y^3$, $v = x + y$, $w = x^2 - xy + y^2$.

15. Small increments. The example which follows is typical of many:

ILLUSTRATION 9. *The sides b, c and the angle A of a triangle ABC are given by*
$$b = 4 \text{ in.}, \quad c = 5 \text{ in.}, \quad A = 30°.$$

A number of children (not knowing these figures) make measurements, with a maximum error of 2 % in length and 3 % in angle. To find the greatest consequent error in their estimates of the area of the triangle.
The area z is given by the formula

$$z = \tfrac{1}{2}bc \sin A.$$

When dealing with products, it is often convenient to begin by taking logarithms, so we write

$$u \equiv \log z = \log \tfrac{1}{2} + \log b + \log c + \log \sin A.$$

Suppose that the errors are δb, δc, δA, leading to an error δz in z or δu in $\log z$. Then we have the approximate formulae

$$\delta u = \frac{du}{dz}\,\delta z,$$

$$\delta u = \frac{\partial u}{\partial b}\,\delta b + \frac{\partial u}{\partial c}\,\delta c + \frac{\partial u}{\partial A}\,\delta A,$$

so that
$$\frac{1}{z}\delta z = \frac{1}{b}\delta b + \frac{1}{c}\delta c + \frac{\cos A}{\sin A}\,\delta A.$$

Now the coefficients $1/b$, $1/c$, $\cos A/\sin A$ are all positive (an important point) and so the greatest error arises when δb, δc, δA are all *positive* and at their greatest values. But we then have $\delta b = 2b/100$, $\delta c = 2c/100$, $\delta A = 3A/100$, so that

$$\frac{1}{z}\delta z = \frac{1}{b}\left(\frac{2b}{100}\right) + \frac{1}{c}\left(\frac{2c}{100}\right) + \frac{\cos A}{\sin A}\left(\frac{3A}{100}\right)$$

$$= \frac{4}{100} + \frac{3A\cot A}{100}.$$

Now the angle A is measured (as in all work for calculus) in *radians*, so that $A = \frac{1}{6}\pi$; also $\cot A = \sqrt{3}$. Hence

$$\frac{1}{z}\delta z = \frac{4}{100} + \frac{\pi\sqrt{3}}{200} = \frac{8+\pi\sqrt{3}}{200},$$

so that
$$\delta z = \frac{(8+\pi\sqrt{3})z}{200}.$$

The greatest percentage error in the estimate of z is thus

$$\tfrac{1}{2}(8+\pi\sqrt{3})\,\% = 6\tfrac{3}{4}\,\%$$

approximately. The corresponding error in actual value is

$$\tfrac{1}{2}\times 4\times 5\times \sin 30°\times \frac{6\tfrac{3}{4}}{100},$$

or
$$0\cdot34\,\text{sq.in.}$$

EXAMPLES IX

Repeat the work of this illustration (with appropriate modifications) for the following sets of measurements:

1. $b = 2$, $c = 3$, $A = 45°$.
2. $b = 4$, $c = 5$, $A = 150°$.
3. $b = 2$, $c = 3$, $A = 135°$.

16. Partial differential coefficients of higher order. If

$$z \equiv f(x, y),$$

then $\dfrac{\partial z}{\partial x}, \dfrac{\partial z}{\partial y}$ are also functions of the two variables x, y, and possess their own differential coefficients. We use notation such as

$$\frac{\partial}{\partial x}\left(\frac{\partial z}{\partial x}\right) \equiv \frac{\partial^2 z}{\partial x^2},$$

$$\frac{\partial}{\partial x}\left(\frac{\partial^2 z}{\partial x^2}\right) \equiv \frac{\partial^3 z}{\partial x^3},$$

$$\frac{\partial}{\partial x}\left(\frac{\partial z}{\partial y}\right) \equiv \frac{\partial^2 z}{\partial x \partial y}, \quad \frac{\partial}{\partial y}\left(\frac{\partial z}{\partial x}\right) \equiv \frac{\partial^2 z}{\partial y \partial x},$$

$$\frac{\partial}{\partial y}\left(\frac{\partial z}{\partial y}\right) \equiv \frac{\partial^2 z}{\partial y^2},$$

and so on.

For example, if

$$z = (ax + by)^3,$$

then

$$\frac{\partial z}{\partial x} = 3a(ax + by)^2,$$

$$\frac{\partial z}{\partial y} = 3b(ax + by)^2,$$

$$\frac{\partial^2 z}{\partial x^2} = 6a^2(ax + by),$$

$$\frac{\partial^2 z}{\partial x \partial y} = 6ab(ax + by) = \frac{\partial^2 z}{\partial y \partial x},$$

$$\frac{\partial^2 z}{\partial y^2} = 6b^2(ax + by).$$

Before going further, we give a proof (under the simplest conditions only) that *the order in which partial differentiations are performed is irrelevant.* In other words, we prove that, in 'normal' cases,

$$\frac{\partial^2 z}{\partial x \partial y} = \frac{\partial^2 z}{\partial y \partial x}.$$

[The proof which follows may be omitted at a first reading.]

Let

$$z \equiv f(x, y)$$

be a given function of the two variables x, y, and consider the expression

$$E \equiv f(x+h, y+k) - f(x+h, y) - f(x, y+k) + f(x, y).$$

First we express E in the form

$$E \equiv \{f(x+h,y+k) - f(x+h,y)\} - \{f(x,y+k) - f(x,y)\},$$

and then write $\quad u(x) \equiv f(x,y+k) - f(x,y),$

where y is regarded as constant. Thus

$$E \equiv u(x+h) - u(x),$$

and we may apply the mean-value theorem (Vol. I, p. 61) for the function $u(x)$ of the single variable x. Hence

$$E \equiv h u'(x+\theta_1 h),$$

where θ_1 lies between 0, 1, the differentiation of u being performed on the assumption that y is constant. Thus, by definition of u,

$$E \equiv h\{f_x'(x+\theta_1 h, y+k) - f_x'(x+\theta_1 h, y)\}.$$

Now write $\quad v(y) \equiv f_x'(x+\theta_1 h, y),$

where $x+\theta_1 h$ is now regarded as constant. Then

$$E \equiv h\{v(y+k) - v(y)\},$$

and we may apply the mean-value theorem for the function $v(y)$ of the single variable y. Hence

$$E \equiv hk v'(y+\theta_2 k),$$

where θ_2 lies between 0, 1, the differentiation of v being performed on the assumption that x is constant. Thus, by definition of v,

$$E \equiv hk f_{yx}''(x+\theta_1 h, y+\theta_2 k),$$

the differentiations of $f(x,y)$ being performed first with respect to x and then with respect to y.

In the same way, we could have written E in the form

$$E \equiv \{f(x+h,y+k) - f(x,y+k)\} - \{f(x+h,y) - f(x,y)\}$$

and so obtained the formula

$$E \equiv kh f_{xy}''(x+\theta_3 h, y+\theta_4 k),$$

where θ_3, θ_4 lie between 0, 1 and the differentiations of $f(x,y)$ are performed first with respect to y and then with respect to x.

Hence $\quad f_{yx}''(x+\theta_1 h, y+\theta_2 k) = f_{xy}''(x+\theta_3 h, y+\theta_4 k).$

Now let $h, k \to 0$ independently. On the assumption that the two second-order partial differential coefficients are both continuous, we reach, as the limiting form of the equation, the formula

$$\frac{\partial^2 f}{\partial y\,\partial x} = \frac{\partial^2 f}{\partial x\,\partial y}.$$

Finally, we remark that, once this result is established, we can easily prove more general results such as

$$\frac{\partial^3 z}{\partial x^2 \partial y} = \frac{\partial^3 z}{\partial x \partial y \partial x} = \frac{\partial^3 z}{\partial y \partial x^2},$$

and so on.

EXAMPLES X

Evaluate $\dfrac{\partial^2 z}{\partial x^2}, \dfrac{\partial^2 z}{\partial x \partial y}, \dfrac{\partial^2 z}{\partial y^2}, \dfrac{\partial^3 z}{\partial x^2 \partial y}, \dfrac{\partial^3 z}{\partial x \partial y^2}$ for each of the following functions:

1. $x^2 y^2$.	2. $x^3 y^4$.	3. $e^x \sin y$.
4. $x \log y$.	5. $x^2 \sin y$.	6. $x \, e^{-y}$.
7. x/y.	8. $y \tan^{-1} x$.	9. $x^2 y^2 e^x$.

ILLUSTRATION 10. *To prove that, if*

$$z = x\phi(y/x) + \psi(y/x),$$

then $\qquad\qquad x^2 \dfrac{\partial^2 z}{\partial x^2} + 2xy \dfrac{\partial^2 z}{\partial x \partial y} + y^2 \dfrac{\partial^2 z}{\partial y^2} = 0,$

where $\phi(y/x)$, $\psi(y/x)$ *are functions of the variable* y/x.

Let ϕ', ψ' be written to denote $\phi'(y/x)$, $\psi'(y/x)$, the differential coefficients of ϕ, ψ with respect to the variable y/x. (In other words, if t is written for y/x, then ϕ' is $\dfrac{d\phi}{dt}$ with y/x written for t *after differentiation*.)

Then $\qquad\qquad \dfrac{\partial}{\partial x}\{\phi(y/x)\} = \phi' \dfrac{\partial}{\partial x}(y/x)$

$$= -\phi' y/x^2,$$

$$\frac{\partial}{\partial y}\{\phi(y/x)\} = \phi' \frac{\partial}{\partial y}(y/x)$$

$$= \phi'/x,$$

with similar results for ψ. Thus, by the definition of z,

$$\frac{\partial z}{\partial x} = \phi - \frac{\phi' y}{x} - \frac{\psi' y}{x^2},$$

$$\frac{\partial z}{\partial y} = \frac{x\phi'}{x} + \frac{\psi'}{x}$$

$$= \phi' + \frac{\psi'}{x}.$$

Hence
$$x\frac{\partial z}{\partial x}+y\frac{\partial z}{\partial y}=x\phi.$$

Differentiate with respect to x, y respectively. Then

$$x\frac{\partial^2 z}{\partial x^2}+\frac{\partial z}{\partial x}+y\frac{\partial^2 z}{\partial x\,\partial y}=\phi+x\frac{\partial\phi}{\partial x}$$

$$=\phi-\phi'y/x;$$

$$x\frac{\partial^2 z}{\partial x\,\partial y}+y\frac{\partial^2 z}{\partial y^2}+\frac{\partial z}{\partial y}=x\frac{\partial\phi}{\partial y}$$

$$=\phi'.$$

Multiply these equations by x, y and add. Then

$$x^2\frac{\partial^2 z}{\partial x^2}+2xy\frac{\partial^2 z}{\partial x\,\partial y}+y^2\frac{\partial^2 z}{\partial y^2}+x\frac{\partial z}{\partial x}+y\frac{\partial z}{\partial y}=x\phi,$$

or, using the previous result,

$$x^2\frac{\partial^2 z}{\partial x^2}+2xy\frac{\partial^2 z}{\partial x\,\partial y}+y^2\frac{\partial^2 z}{\partial y^2}=0.$$

ILLUSTRATION 11. *(Change of variable.) To prove that, if*

$$\frac{\partial^2 z}{\partial x^2}=\frac{\partial^2 z}{\partial y^2},$$

then the function z must be of the form

$$z=f(x+y)+g(x-y),$$

where f, g are arbitrary functions of their arguments.

Write
$$\xi=x+y,$$
$$\eta=x-y.$$

Then
$$\frac{\partial z}{\partial x}=\frac{\partial z}{\partial\xi}\frac{\partial\xi}{\partial x}+\frac{\partial z}{\partial\eta}\frac{\partial\eta}{\partial x}$$

$$=\frac{\partial z}{\partial\xi}+\frac{\partial z}{\partial\eta},$$

and
$$\frac{\partial z}{\partial y}=\frac{\partial z}{\partial\xi}\frac{\partial\xi}{\partial y}+\frac{\partial z}{\partial\eta}\frac{\partial\eta}{\partial y}$$

$$=\frac{\partial z}{\partial\xi}-\frac{\partial z}{\partial\eta}.$$

It follows that the 'operation'

$$\frac{\partial}{\partial x}$$

on a function of x, y is equivalent to the 'operation'

$$\frac{\partial}{\partial \xi} + \frac{\partial}{\partial \eta}$$

on the same function expressed in terms of ξ, η; and that

$$\frac{\partial}{\partial y}$$

is equivalent to $\quad \dfrac{\partial}{\partial \xi} - \dfrac{\partial}{\partial \eta}.$

Thus $\quad \dfrac{\partial^2 z}{\partial x^2} = \dfrac{\partial}{\partial x}\left(\dfrac{\partial z}{\partial x}\right) \qquad\qquad$ (by definition)

$$= \frac{\partial}{\partial x}\left(\frac{\partial z}{\partial \xi} + \frac{\partial z}{\partial \eta}\right) \qquad\qquad \left(\text{replacing } \frac{\partial z}{\partial x}\right)$$

$$= \left(\frac{\partial}{\partial \xi} + \frac{\partial}{\partial \eta}\right)\left(\frac{\partial z}{\partial \xi} + \frac{\partial z}{\partial \eta}\right) \qquad \left(\text{replacing } \frac{\partial}{\partial x}\right)$$

$$= \frac{\partial}{\partial \xi}\left(\frac{\partial z}{\partial \xi} + \frac{\partial z}{\partial \eta}\right) + \frac{\partial}{\partial \eta}\left(\frac{\partial z}{\partial \xi} + \frac{\partial z}{\partial \eta}\right)$$

$$= \frac{\partial^2 z}{\partial \xi^2} + 2\frac{\partial^2 z}{\partial \xi \partial \eta} + \frac{\partial^2 z}{\partial \eta^2};$$

and, by similar argument,

$$\frac{\partial^2 z}{\partial y^2} = \frac{\partial^2 z}{\partial \xi^2} - 2\frac{\partial^2 z}{\partial \xi \partial \eta} + \frac{\partial^2 z}{\partial \eta^2}.$$

Hence the given relation

$$\frac{\partial^2 z}{\partial x^2} = \frac{\partial^2 z}{\partial y^2}$$

is equivalent to $\quad \dfrac{\partial^2 z}{\partial \xi \partial \eta} = 0.$

Thus $\quad \dfrac{\partial}{\partial \xi}\left(\dfrac{\partial z}{\partial \eta}\right) = 0,$

so that $\dfrac{\partial z}{\partial \eta}$ is independent of ξ and therefore a function of η only; say

$$\frac{\partial z}{\partial \eta} = u(\eta).$$

Integrating, $\quad z = \displaystyle\int u(\eta)\, d\eta + A,$

where A is constant *with respect to* η, but may involve ξ.

Writing $$\int u(\eta)\,d\eta = g(\eta)$$

and $$A = f(\xi),$$

we have the relation $\quad z = f(\xi) + g(\eta),$

or, in terms of x, y, $\quad z = f(x+y) + g(x-y).$

ILLUSTRATION 12. (*Change of variable by the use of operators.*) *Given that*

$$u = x^2 - y^2, \quad v = 2xy,$$

to express

$$\frac{1}{x^2+y^2}\left(\frac{\partial^2\phi}{\partial x^2} + \frac{\partial^2\phi}{\partial y^2}\right)$$

with u, v as independent variables.

This example is typical of many, and is given in detail. (The beginner may also like to solve it by straightforward partial differentiations.) We assume throughout that the context makes it clear when ϕ is being regarded as a function of x, y and when it is being regarded as a function of u, v.

By the chain rule of § 10 (p. 25)

$$\frac{\partial\phi}{\partial x} = \frac{\partial\phi}{\partial u}\frac{\partial u}{\partial x} + \frac{\partial\phi}{\partial v}\frac{\partial v}{\partial x}$$

$$= \frac{\partial\phi}{\partial u}(2x) + \frac{\partial\phi}{\partial v}(2y);$$

and $$\frac{\partial\phi}{\partial y} = \frac{\partial\phi}{\partial u}\frac{\partial u}{\partial y} + \frac{\partial\phi}{\partial v}\frac{\partial v}{\partial y}$$

$$= \frac{\partial\phi}{\partial u}(-2y) + \frac{\partial\phi}{\partial v}(2x).$$

Hence $$x\frac{\partial\phi}{\partial x} + y\frac{\partial\phi}{\partial y} = 2(x^2-y^2)\frac{\partial\phi}{\partial u} + 4xy\frac{\partial\phi}{\partial v}$$

$$= 2u\frac{\partial\phi}{\partial u} + 2v\frac{\partial\phi}{\partial v};$$

and $$y\frac{\partial\phi}{\partial x} - x\frac{\partial\phi}{\partial y} = 4xy\frac{\partial\phi}{\partial u} + 2(y^2-x^2)\frac{\partial\phi}{\partial v}$$

$$= 2v\frac{\partial\phi}{\partial u} - 2u\frac{\partial\phi}{\partial v}.$$

We may regard the symbol

$$x\frac{\partial}{\partial x} + y\frac{\partial}{\partial y}$$

as an OPERATOR in the sense that, when it 'operates' on the function ϕ (expressed in terms of x, y), the result is

$$x\frac{\partial \phi}{\partial x} + y\frac{\partial \phi}{\partial y}.$$

What we have just proved is that the operators

$$x\frac{\partial}{\partial x} + y\frac{\partial}{\partial y}, \quad y\frac{\partial}{\partial x} - x\frac{\partial}{\partial y},$$

acting on any function of x, y, are equivalent to the operators

$$2u\frac{\partial}{\partial u} + 2v\frac{\partial}{\partial v}, \quad 2v\frac{\partial}{\partial u} - 2u\frac{\partial}{\partial v}$$

operating on that function when expressed in terms of u, v.

[The next steps may be abbreviated with experience, but are given in full for clarity.]

In particular,

$$\left(x\frac{\partial}{\partial x} + y\frac{\partial}{\partial y}\right)\left(x\frac{\partial \phi}{\partial x} + y\frac{\partial \phi}{\partial y}\right)$$

$$\equiv \left(2u\frac{\partial}{\partial u} + 2v\frac{\partial}{\partial v}\right)\left(2u\frac{\partial \phi}{\partial u} + 2v\frac{\partial \phi}{\partial v}\right),$$

since the functions $x\frac{\partial \phi}{\partial x} + y\frac{\partial \phi}{\partial y}$, $2u\frac{\partial \phi}{\partial u} + 2v\frac{\partial \phi}{\partial v}$ and the operators $\left(x\frac{\partial}{\partial x} + y\frac{\partial}{\partial y}\right)$, $\left(2u\frac{\partial}{\partial u} + 2v\frac{\partial}{\partial v}\right)$ are respectively equivalent.

Consider the left-hand side.

[This step should be noted carefully, as it is often calculated wrongly.]

This is equal to

$$\left(x\frac{\partial}{\partial x} + y\frac{\partial}{\partial y}\right)\left(x\frac{\partial \phi}{\partial x}\right) + \left(x\frac{\partial}{\partial x} + y\frac{\partial}{\partial y}\right)\left(y\frac{\partial \phi}{\partial y}\right),$$

or

$$x\frac{\partial}{\partial x}\left(x\frac{\partial \phi}{\partial x}\right)$$

$$+ y\frac{\partial}{\partial y}\left(x\frac{\partial \phi}{\partial x}\right)$$

$$+ x\frac{\partial}{\partial x}\left(y\frac{\partial \phi}{\partial y}\right)$$

$$+ y\frac{\partial}{\partial y}\left(y\frac{\partial \phi}{\partial y}\right),$$

or
$$x\left(x\frac{\partial^2\phi}{\partial x^2}+\frac{\partial\phi}{\partial x}\right)$$
$$+y\left(x\frac{\partial^2\phi}{\partial y\,\partial x}\right)$$
$$+x\left(y\frac{\partial^2\phi}{\partial x\,\partial y}\right)$$
$$+y\left(y\frac{\partial^2\phi}{\partial y^2}+\frac{\partial\phi}{\partial y}\right),$$

or, finally, $x^2\dfrac{\partial^2\phi}{\partial x^2}+2xy\dfrac{\partial^2\phi}{\partial x\,\partial y}+y^2\dfrac{\partial^2\phi}{\partial y^2}+x\dfrac{\partial\phi}{\partial x}+y\dfrac{\partial\phi}{\partial y}.$

Hence we have the equation

$$x^2\frac{\partial^2\phi}{\partial x^2}+2xy\frac{\partial^2\phi}{\partial x\,\partial y}+y^2\frac{\partial^2\phi}{\partial y^2}+x\frac{\partial\phi}{\partial x}+y\frac{\partial\phi}{\partial y}$$
$$=4\left(u^2\frac{\partial^2\phi}{\partial u^2}+2uv\frac{\partial^2\phi}{\partial u\,\partial v}+v^2\frac{\partial^2\phi}{\partial v^2}+u\frac{\partial\phi}{\partial u}+v\frac{\partial\phi}{\partial v}\right).$$

Now consider similarly the relation

$$\left(y\frac{\partial}{\partial x}-x\frac{\partial}{\partial y}\right)\left(y\frac{\partial\phi}{\partial x}-x\frac{\partial\phi}{\partial y}\right)$$
$$\equiv\left(2v\frac{\partial}{\partial u}-2u\frac{\partial}{\partial v}\right)\left(2v\frac{\partial\phi}{\partial u}-2u\frac{\partial\phi}{\partial v}\right).$$

The left-hand side is

$$y\frac{\partial}{\partial x}\left(y\frac{\partial\phi}{\partial x}\right)$$
$$-x\frac{\partial}{\partial y}\left(y\frac{\partial\phi}{\partial x}\right)$$
$$-y\frac{\partial}{\partial x}\left(x\frac{\partial\phi}{\partial y}\right)$$
$$+x\frac{\partial}{\partial y}\left(x\frac{\partial\phi}{\partial y}\right),$$

or
$$y\left(y\frac{\partial^2\phi}{\partial x^2}\right)$$
$$-x\left(y\frac{\partial^2\phi}{\partial y\,\partial x}+\frac{\partial\phi}{\partial x}\right)$$
$$-y\left(x\frac{\partial^2\phi}{\partial x\,\partial y}+\frac{\partial\phi}{\partial y}\right)$$
$$+x\left(x\frac{\partial^2\phi}{\partial y^2}\right).$$

Hence

$$y^2 \frac{\partial^2 \phi}{\partial x^2} - 2xy \frac{\partial^2 \phi}{\partial x \partial y} + x^2 \frac{\partial^2 \phi}{\partial y^2} - x \frac{\partial \phi}{\partial x} - y \frac{\partial \phi}{\partial y}$$

$$= 4\left(v^2 \frac{\partial^2 \phi}{\partial u^2} - 2uv \frac{\partial^2 \phi}{\partial u \partial v} + u^2 \frac{\partial^2 \phi}{\partial v^2} - u \frac{\partial \phi}{\partial u} - v \frac{\partial \phi}{\partial v}\right).$$

Adding this to the similar equation obtained above, we have

$$(x^2 + y^2)\left(\frac{\partial^2 \phi}{\partial x^2} + \frac{\partial^2 \phi}{\partial y^2}\right) = 4(u^2 + v^2)\left(\frac{\partial^2 \phi}{\partial u^2} + \frac{\partial^2 \phi}{\partial v^2}\right).$$

But $$u^2 + v^2 = (x^2 - y^2)^2 + (2xy)^2 = (x^2 + y^2)^2,$$

so that $$\frac{1}{x^2 + y^2}\left(\frac{\partial^2 \phi}{\partial x^2} + \frac{\partial^2 \phi}{\partial y^2}\right) = 4\left(\frac{\partial^2 \phi}{\partial u^2} + \frac{\partial^2 \phi}{\partial v^2}\right).$$

ILLUSTRATION 13. *The expression $\dfrac{\partial^2 \phi}{\partial x^2} + \dfrac{\partial^2 \phi}{\partial y^2}$ in polar coordinates.*
The expression

$$\frac{\partial^2 \phi}{\partial x^2} + \frac{\partial^2 \phi}{\partial y^2}$$

is of great importance in mathematical physics, and it is important to be able to transform it to polar coordinates in accordance with the usual formulae
$$x = r \cos \theta, \quad y = r \sin \theta.$$

We use the method of the preceding illustration, now expressing the argument with more normal brevity.

By the chain rule of § 10 (p. 25),

$$\frac{\partial \phi}{\partial r} = \frac{\partial \phi}{\partial x}\frac{\partial x}{\partial r} + \frac{\partial \phi}{\partial y}\frac{\partial y}{\partial r}$$

$$= \frac{\partial \phi}{\partial x} \cos \theta + \frac{\partial \phi}{\partial y} \sin \theta,$$

so that $$r \frac{\partial \phi}{\partial r} = x \frac{\partial \phi}{\partial x} + y \frac{\partial \phi}{\partial y}.$$

Hence $$r \frac{\partial}{\partial r}\left(r \frac{\partial \phi}{\partial r}\right) = \left(x \frac{\partial}{\partial x} + y \frac{\partial}{\partial y}\right)\left(x \frac{\partial \phi}{\partial x} + y \frac{\partial \phi}{\partial y}\right),$$

or $$r^2 \frac{\partial^2 \phi}{\partial r^2} + r \frac{\partial \phi}{\partial r} = x^2 \frac{\partial^2 \phi}{\partial x^2} + 2xy \frac{\partial^2 \phi}{\partial x \partial y} + y^2 \frac{\partial^2 \phi}{\partial y^2} + x \frac{\partial \phi}{\partial x} + y \frac{\partial \phi}{\partial y}.$$

Also
$$\frac{\partial \phi}{\partial \theta} = \frac{\partial \phi}{\partial x}\frac{\partial x}{\partial \theta} + \frac{\partial \phi}{\partial y}\frac{\partial y}{\partial \theta}$$

$$= \frac{\partial \phi}{\partial x}(-r\sin\theta) + \frac{\partial \phi}{\partial y}(r\cos\theta)$$

$$= -y\frac{\partial \phi}{\partial x} + x\frac{\partial \phi}{\partial y}.$$

Hence
$$\frac{\partial^2 \phi}{\partial \theta^2} = y^2\frac{\partial^2 \phi}{\partial x^2} - 2xy\frac{\partial^2 \phi}{\partial x \partial y} + x^2\frac{\partial^2 \phi}{\partial y^2} - x\frac{\partial \phi}{\partial x} - y\frac{\partial \phi}{\partial y}.$$

Adding,
$$r^2\frac{\partial^2 \phi}{\partial r^2} + r\frac{\partial \phi}{\partial r} + \frac{\partial^2 \phi}{\partial \theta^2} = (x^2+y^2)\left(\frac{\partial^2 \phi}{\partial x^2} + \frac{\partial^2 \phi}{\partial y^2}\right)$$

$$= r^2\left(\frac{\partial^2 \phi}{\partial x^2} + \frac{\partial^2 \phi}{\partial y^2}\right),$$

so that
$$\frac{\partial^2 \phi}{\partial x^2} + \frac{\partial^2 \phi}{\partial y^2} = \frac{\partial^2 \phi}{\partial r^2} + \frac{1}{r}\frac{\partial \phi}{\partial r} + \frac{1}{r^2}\frac{\partial^2 \phi}{\partial \theta^2}.$$

ILLUSTRATION 14. *Homogeneous functions.* A function of three variables $f(x, y, z)$ is said to be *homogeneous of degree n* if it possesses the property
$$f(ux, uy, uz) \equiv u^n f(x, y, z)$$

for some number n. Typical examples are

$$\frac{x^2+y^2+z^2}{x-y+z} \quad (n=1),$$

$$e^{(x+y)/z} \quad (n=0),$$

$$\frac{1}{x^3+2y^3+3z^3} \quad (n=-3).$$

Another example is $\sqrt{x}+\sqrt{y}+\sqrt{z} \quad (n=\frac{1}{2})$,

where it is assumed (for real functions) that x, y, z are positive, and that u also is positive.

We prove the following theorem:

If $$f(x, y, z)$$

is a function homogeneous of degree n in x, y, z, then

$$x\frac{\partial f}{\partial x} + y\frac{\partial f}{\partial y} + z\frac{\partial f}{\partial z} \equiv nf.$$

Suppose that x, y, z are temporarily fixed. Then

$$f(ux, uy, uz)$$

is a function of the single variable u. Write

$$ux = \xi, \quad uy = \eta, \quad uz = \zeta.$$

Then, in accordance with the rule for differentiation (compare Illustration 5 on p. 24),

$$\frac{d}{du} f(ux, uy, uz)$$

$$\equiv \frac{\partial f(\xi, \eta, \zeta)}{\partial \xi} \frac{d\xi}{du} + \frac{\partial f(\xi, \eta, \zeta)}{\partial \eta} \frac{d\eta}{du} + \frac{\partial f(\xi, \eta, \zeta)}{\partial \zeta} \frac{d\zeta}{du}$$

$$\equiv x \frac{\partial f(\xi, \eta, \zeta)}{\partial \xi} + y \frac{\partial f(\xi, \eta, \zeta)}{\partial \eta} + z \frac{\partial f(\xi, \eta, \zeta)}{\partial \zeta}.$$

But, by definition,

$$f(ux, uy, uz) \equiv u^n f(x, y, z),$$

so that $(x, y, z$ still being fixed) we also have the relation

$$\frac{d}{du} f(ux, uy, uz) \equiv n u^{n-1} f(x, y, z).$$

Hence, equating values of $\dfrac{d}{du} f(ux, uy, uz)$,

$$x \frac{\partial f(\xi, \eta, \zeta)}{\partial \xi} + y \frac{\partial f(\xi, \eta, \zeta)}{\partial \eta} + z \frac{\partial f(\xi, \eta, \zeta)}{\partial \zeta} \equiv n u^{n-1} f(x, y, z).$$

In particular, this result is true for $u = 1$; and then ξ, η, ζ are respectively equal to $x, y\ z$. Hence

$$x \frac{\partial f(x, y, z)}{\partial x} + y \frac{\partial f(x, y, z)}{\partial y} + z \frac{\partial f(x, y, z)}{\partial z} = n f(x, y, z),$$

or, more briefly, $\qquad x \dfrac{\partial f}{\partial x} + y \dfrac{\partial f}{\partial y} + z \dfrac{\partial f}{\partial z} = nf.$

This result may be extended. By differentiating with respect to x, y, z respectively, we obtain the relations

$$x \frac{\partial^2 f}{\partial x^2} + \frac{\partial f}{\partial x} + y \frac{\partial^2 f}{\partial x \partial y} + z \frac{\partial^2 f}{\partial z \partial x} = n \frac{\partial f}{\partial x},$$

$$x \frac{\partial^2 f}{\partial x \partial y} + y \frac{\partial^2 f}{\partial y^2} + \frac{\partial f}{\partial y} + z \frac{\partial^2 f}{\partial y \partial z} = n \frac{\partial f}{\partial y}.$$

$$x \frac{\partial^2 f}{\partial z \partial x} + y \frac{\partial^2 f}{\partial y \partial z} + z \frac{\partial^2 f}{\partial z^2} + \frac{\partial f}{\partial z} = n \frac{\partial f}{\partial z}.$$

Multiply by x, y, z in turn and add, first transferring the terms in $\dfrac{\partial f}{\partial x}, \dfrac{\partial f}{\partial y}, \dfrac{\partial f}{\partial z}$ to the right-hand side. Then

$$x^2\frac{\partial^2 f}{\partial x^2} + y^2\frac{\partial^2 f}{\partial y^2} + z^2\frac{\partial^2 f}{\partial z^2} + 2yz\frac{\partial^2 f}{\partial y\,\partial z} + 2zx\frac{\partial^2 f}{\partial z\,\partial x} + 2xy\frac{\partial^2 f}{\partial x\,\partial y}$$

$$= (n-1)\left\{x\frac{\partial f}{\partial x} + y\frac{\partial f}{\partial y} + z\frac{\partial f}{\partial z}\right\}$$

$$= n(n-1)f.$$

17. Implicit functions of a single variable. A function y of a variable x is often defined, not directly in the form $y = f(x)$, but implicitly by means of an equation

$$f(x, y) = 0.$$

The differential coefficients $\dfrac{dy}{dx}, \dfrac{d^2y}{dx^2}$ of y with respect to x may then be calculated as follows:

Since x, y are both functions of x, we have, by the chain rule, the relation

$$\frac{df}{dx} = \frac{\partial f}{\partial x} + \frac{\partial f}{\partial y}\frac{dy}{dx},$$

so that, since f is always zero,

$$0 = \frac{\partial f}{\partial x} + \frac{\partial f}{\partial y}\frac{dy}{dx}.$$

Hence

$$\frac{dy}{dx} = -\frac{\dfrac{\partial f}{\partial x}}{\dfrac{\partial f}{\partial y}}.$$

Moreover, applying the same treatment to $\dfrac{\partial f}{\partial x}, \dfrac{\partial f}{\partial y}$ respectively,

$$\frac{d}{dx}\left(\frac{\partial f}{\partial x}\right) = \frac{\partial^2 f}{\partial x^2} + \frac{\partial^2 f}{\partial x\,\partial y}\frac{dy}{dx},$$

$$\frac{d}{dx}\left(\frac{\partial f}{\partial y}\right) = \frac{\partial^2 f}{\partial x\,\partial y} + \frac{\partial^2 f}{\partial y^2}\frac{dy}{dx},$$

so that, on differentiating the relation

$$0 = \frac{\partial f}{\partial x} + \frac{\partial f}{\partial y}\frac{dy}{dx}$$

with respect to x,

$$\left(\frac{\partial^2 f}{\partial x^2}+\frac{\partial^2 f}{\partial x\,\partial y}\frac{dy}{dx}\right)+\left(\frac{\partial^2 f}{\partial x\,\partial y}+\frac{\partial^2 f}{\partial y^2}\frac{dy}{dx}\right)\frac{dy}{dx}+\frac{\partial f}{\partial y}\frac{d^2 y}{dx^2}=0,$$

or

$$\frac{\partial^2 f}{\partial x^2}+2\frac{\partial^2 f}{\partial x\,\partial y}\frac{dy}{dx}+\frac{\partial^2 f}{\partial y^2}\left(\frac{dy}{dx}\right)^2+\frac{\partial f}{\partial y}\frac{d^2 y}{dx^2}=0.$$

Hence

$$\frac{d^2 y}{dx^2}=-\frac{\left(\dfrac{\partial f}{\partial y}\right)^2\dfrac{\partial^2 f}{\partial x^2}-2\dfrac{\partial f}{\partial x}\dfrac{\partial f}{\partial y}\dfrac{\partial^2 f}{\partial x\,\partial y}+\left(\dfrac{\partial f}{\partial x}\right)^2\dfrac{\partial^2 f}{\partial y^2}}{\left(\dfrac{\partial f}{\partial y}\right)^3}.$$

ILLUSTRATION 15. *To find expressions for* $\dfrac{dy}{dx},\dfrac{d^2 y}{dx^2}$ *when*

$$x^2+y^2-1=0.$$

Differentiating, we have

$$x+y\frac{dy}{dx}=0,$$

so that

$$\frac{dy}{dx}=-\frac{x}{y}.$$

Differentiating again,

$$1+\left(\frac{dy}{dx}\right)^2+y\frac{d^2 y}{dx^2}=0,$$

or

$$1+\frac{x^2}{y^2}+y\frac{d^2 y}{dx^2}=0,$$

or

$$y^2+x^2+y^3\frac{d^2 y}{dx^2}=0.$$

Hence

$$\frac{d^2 y}{dx^2}=-\frac{1}{y^3}.$$

EXAMPLES XI

Find expressions for $\dfrac{dy}{dx},\dfrac{d^2 y}{dx^2}$ in each of the following cases:

1. $y^2-4ax+a^2=0.$ 2. $\alpha x^2+\beta y^2-1=0.$
3. $x^3+y^3=a^3.$ 4. $xy^2+yx^2=1.$

18. The tangent to the curve $f(x,y)=0$. Let (p,q) be a given point lying on the curve whose equation is

$$f(x,y)=0.$$

The gradient at the point is $\dfrac{dy}{dx}$, where

$$\frac{dy}{dx} = -\frac{\dfrac{\partial f}{\partial x}}{\dfrac{\partial f}{\partial y}}$$

(p. 49), the differential coefficients being evaluated when $x = p$ $y = q$. Hence the equation of the tangent is

$$y - q = -\frac{\dfrac{\partial f}{\partial x}}{\dfrac{\partial f}{\partial y}} (x - p),$$

or $$\left[\frac{\partial f}{\partial x}\right]_{(p, q)} (x - p) + \left[\frac{\partial f}{\partial y}\right]_{(p, q)} (y - q) = 0,$$

the square brackets being inserted to imply evaluation with $x = p, y = q$.

ILLUSTRATION 16. *To find the equation of the tangent to the ellipse*

$$\frac{x^2}{a^2} + \frac{y^2}{b^2} - = 1 \, 0$$

at the point (p, q).

We have $\dfrac{\partial f}{\partial x} = \dfrac{2x}{a^2}, \quad \dfrac{\partial f}{\partial y} = \dfrac{2y}{b^2},$

so that $\left[\dfrac{\partial f}{\partial x}\right]_{(p, q)} = \dfrac{2p}{a^2}, \quad \left[\dfrac{\partial f}{\partial y}\right]_{(p, q)} = \dfrac{2q}{b^2}.$

Hence the equation of the tangent is

$$\frac{2p}{a^2} (x - p) + \frac{2q}{b^2} (y - q) = 0,$$

or, since $p^2/a^2 + q^2/b^2 = 1$, $\dfrac{px}{a^2} + \dfrac{qy}{b^2} = 1.$

19. Exact differentials. We are now familiar with the formula $dz = \dfrac{\partial z}{\partial x} dx + \dfrac{\partial z}{\partial y} dy$ for the differential of a function z of two variables x, y; but the converse problem often arises and requires an answer:

To determine whether, when two functions $f(x, y)$, $g(x, y)$ are given, there exists a function z of x, y whose differential is given by the relation

$$dz = f(x, y) \, dx + g(x, y) \, dy.$$

If z does exist, then, necessarily,

$$\frac{\partial z}{\partial x} = f(x, y), \quad \frac{\partial z}{\partial y} = g(x, y),$$

and these two partial differential coefficients satisfy (for normal conditions) the relation (p. 38)

$$\frac{\partial}{\partial y}\left(\frac{\partial z}{\partial x}\right) = \frac{\partial}{\partial x}\left(\frac{\partial z}{\partial y}\right).$$

Hence a NECESSARY condition for the existence of the function z is that

$$\frac{\partial}{\partial y}\{f(x, y)\} = \frac{\partial}{\partial x}\{g(x, y)\}.$$

The condition is also SUFFICIENT:
If two functions $f \equiv f(x, y)$, $g \equiv g(x, y)$ are such that

$$\frac{\partial f}{\partial y} = \frac{\partial g}{\partial x},$$

then there exists a function z of x, y such that

$$dz = f\,dx + g\,dy.$$

[This proof may be omitted at a first reading.]

Take the function $f(x, y)$ and integrate it on the assumption that y is to be kept constant; this gives a function

$$F(x, y) \equiv \int_{(y \text{ const.})} f(x, y)\,dx.$$

In the same way, integration of $g(x, y)$ on the assumption that x is to be kept constant gives the function

$$G(x, y) = \int_{(x \text{ const.})} g(x, y)\,dy.$$

The functions F, G satisfy the relations (Vol. I, pp. 86–7)

$$\frac{\partial F}{\partial x} = f,$$

$$\frac{\partial G}{\partial y} = g,$$

so that, in virtue of the given relation $\dfrac{\partial f}{\partial y} = \dfrac{\partial g}{\partial x}$, we have

$$\frac{\partial^2 F}{\partial y\,\partial x} = \frac{\partial^2 G}{\partial x\,\partial y}.$$

If we assume that F is a function of 'normal' type for which $\dfrac{\partial^2 F}{\partial y \, \partial x} = \dfrac{\partial^2 F}{\partial x \, \partial y}$, then

$$\frac{\partial^2 F}{\partial x \, \partial y} = \frac{\partial^2 G}{\partial x \, \partial y},$$

or

$$\frac{\partial^2 (F - G)}{\partial x \, \partial y} = 0.$$

Hence (compare pp. 42–3) the difference $F - G$ is of the form

$$F - G \equiv u(x) - v(y),$$

where $u(x)$, $v(y)$ are functions of x, y only. It follows that

$$F(x, y) + v(y) \equiv G(x, y) + u(x),$$

and we take either of these expressions as the function z whose differential we require, namely,

$$z \equiv F(x, y) + v(y) \equiv G(x, y) + u(x).$$

With this choice of z, we have

$$\frac{\partial z}{\partial x} = \frac{\partial F}{\partial x} = f,$$

$$\frac{\partial z}{\partial y} = \frac{\partial G}{\partial y} = g,$$

so that

$$f \, dx + g \, dy = \frac{\partial z}{\partial x} \, dx + \frac{\partial z}{\partial y} \, dy$$

$$= dz.$$

DEFINITION. *When the condition*

$$\frac{\partial}{\partial y} \{ f(x, y) \} = \frac{\partial}{\partial x} \{ g(x, y) \}$$

is satisfied, the expression

$$f(x, y) \, dx + g(x, y) \, dy$$

is called an EXACT DIFFERENTIAL.

ILLUSTRATION 17. The steps of the proof may be grasped more clearly by reference to a particular example, say

$$3x^2 (x + y) \, dx + (x^3 + 3y) \, dy.$$

Integrating the first part with respect to x, on the assumption that y is constant, we have

$$F(x, y) \equiv \tfrac{3}{4} x^4 + x^3 y,$$

and integrating the second part with respect to y, on the assumption that x is constant, we have

$$G(x, y) \equiv x^3 y + \tfrac{3}{2} y^2.$$

What, in essence, we have to do is to add to $F(x, y)$ a function of y, and to $G(x, y)$ a function of x, so as to yield (if possible) the same function z. This we do by taking

$$z \equiv \tfrac{3}{4} x^4 + x^3 y + \tfrac{3}{2} y^2.$$

Then $\qquad dz = (3x^3 + 3x^2 y)\,dx + (x^3 + 3y)\,dy,$

as required.

The point of the proof given above is that, since

$$\frac{\partial}{\partial x}(x^3 + 3y) = \frac{\partial}{\partial y}(3x^3 + 3x^2 y),$$

functions $u(x)$, $v(y)$ *must* exist such that

$$F(x, y) + v(y) \equiv G(x, y) + u(x).$$

When they have been determined, either side of this identity gives a function z.

20. Integrating factor. Although there is not, in general, a function z whose differential dz is equal to the given expression

$$f\,dx + g\,dy,$$

it nevertheless happens in several important instances that a function

$$\mu \equiv \mu(x, y)$$

exists with the property that the product

$$\mu(f\,dx + g\,dy)$$

is an exact differential. The function $\mu(x, y)$ is then called an INTEGRATING FACTOR of the given expression.

The condition given in the preceding paragraph becomes

$$\frac{\partial}{\partial y}(\mu f) = \frac{\partial}{\partial x}(\mu g),$$

or $\qquad\qquad \mu\dfrac{\partial f}{\partial y} + f\dfrac{\partial \mu}{\partial y} = \mu\dfrac{\partial g}{\partial x} + g\dfrac{\partial \mu}{\partial x},$

or $\qquad\qquad g\dfrac{\partial \mu}{\partial x} - f\dfrac{\partial \mu}{\partial y} = \mu\left(\dfrac{\partial f}{\partial y} - \dfrac{\partial g}{\partial x}\right).$

This equation, which we may expect to be of help in deriving μ, is unfortunately somewhat awkward to handle; but in practice what we often want is a solution in which μ is a function of x only, in which case the working becomes more manageable:

If μ is a function of x only, then

$$\frac{\partial \mu}{\partial y} = 0$$

and $\dfrac{\partial \mu}{\partial x}$ is the ordinary differential coefficient $\dfrac{d\mu}{dx}$. Then

$$g\frac{d\mu}{dx} = \mu\left(\frac{\partial f}{\partial y} - \frac{\partial g}{\partial x}\right),$$

so that

$$\frac{d\mu}{\mu} = \frac{\dfrac{\partial f}{\partial y} - \dfrac{\partial g}{\partial x}}{g}dx.$$

If it happens that the coefficient of dx is a function of x only, the function $\mu(x)$ can be found; for then

$$\log \mu = \int \frac{\dfrac{\partial f}{\partial y} - \dfrac{\partial g}{\partial x}}{g}dx.$$

The Illustration which follows shows how an integrating factor may be found—though an experienced mathematician will usually 'spot' the solution in such a direct example without going through the routine process.

ILLUSTRATION 18. *To find the current y at time t in an electric circuit with self-inductance L and resistance R under an electromotive force $E \cos pt$, where L, R, E, p are constants.*

It is known that y satisfies the differential equation

$$L\frac{dy}{dt} + Ry = E \cos pt,$$

or, on multiplying by the differential dt,

$$L\,dy + Ry\,dt = E \cos pt\,dt.$$

We seek to find whether the left-hand side is an exact differential. If so,

$$\frac{\partial}{\partial t}(L) = \frac{\partial}{\partial y}(Ry),$$

or

$$0 = R,$$

which is not true in general.

We next try to find an integrating factor μ, by means of which the equation will become

$$\mu L\,dy + \mu Ry\,dt = \mu E \cos pt\,dt.$$

The right-hand side can be integrated (perhaps with difficulty, of course) if μ is a function of t only. Let us therefore try to make the left-hand side an exact differential on that assumption. We must have

$$\frac{\partial}{\partial t}(\mu L) = \frac{\partial}{\partial y}(\mu Ry),$$

or, since μ is a function of t only,

$$L\frac{d\mu}{dt} = \mu R,$$

or

$$\frac{d\mu}{\mu} = \frac{R}{L}dt.$$

Hence

$$\log \mu = Rt/L,$$

or

$$\mu = e^{Rt/L},$$

no constant being introduced since any one particular value of μ will suffice.

The equation is therefore

$$L\,e^{Rt/L}\,dy + Ry\,e^{Rt/L}\,dt = E\,e^{Rt/L} \cos pt\,dt,$$

or

$$L\,d(y\,e^{Rt/L}) = E\,e^{Rt/L} \cos pt\,dt,$$

or

$$y\,e^{Rt/L} = (E/L)\int e^{Rt/L} \cos pt\,dt + C,$$

where C is an arbitrary constant. Evaluating the integral by parts in the usual way, we easily reach the solution

$$y\,e^{Rt/L} = \frac{E\,e^{Rt/L}}{R^2 + L^2 p^2}\{R \cos pt + Lp \sin pt\} + C.$$

The current y at time t is therefore given by the formula

$$y = \frac{E}{R^2 + L^2 p^2}\{R \cos pt + Lp \sin pt\} + C\,e^{-Rt/L},$$

where C is a constant whose value, for a given problem, depends on the initial conditions.

21. Taylor's theorem for several variables. In Vol. II, p. 44, we discussed Taylor's theorem for functions of a single variable. In enunciating the extension we have to use two points of notation which we shall explain almost immediately:

TAYLOR'S THEOREM: To prove that, *if* $f(x, y, z)$ *is a function of the three variables* x, y, z, *then a number* θ, *lying between* 0, 1, *can be found such that*

$$f(x+h, y+j, z+k)$$

$$= f(x, y, z)$$

$$+ \left(h\frac{\partial}{\partial x} + j\frac{\partial}{\partial y} + k\frac{\partial}{\partial z} \right) f(x, y, z)$$

$$+ \frac{1}{2!} \left(h\frac{\partial}{\partial x} + j\frac{\partial}{\partial y} + k\frac{\partial}{\partial z} \right)^2 f(x, y, z)$$

$$\dots\dots\dots\dots\dots\dots\dots\dots\dots\dots\dots$$

$$+ \frac{1}{(n-1)!} \left(h\frac{\partial}{\partial x} + j\frac{\partial}{\partial y} + k\frac{\partial}{\partial z} \right)^{n-1} f(x, y, z)$$

$$+ \frac{1}{n!} \left(h\frac{\partial}{\partial x} + j\frac{\partial}{\partial y} + k\frac{\partial}{\partial z} \right)^n f(x+\theta h, y+\theta j, z+\theta k).$$

It is assumed that the function and all its partial differential co-efficients exist and are continuous for the ranges of values to be considered.

[For the 'shape' of this expression, note the analogy with the binomial theorem.]

NOTATION. By the expression

$$\left(h\frac{\partial}{\partial x} + j\frac{\partial}{\partial y} + k\frac{\partial}{\partial z} \right)^p f(x, y, z)$$

we mean the result of operating p times in succession on the function $f(x, y, z)$ by the operator $h\frac{\partial}{\partial x} + j\frac{\partial}{\partial y} + k\frac{\partial}{\partial z}$; for example, when $p = 1, 2$ we have respectively

$$h\frac{\partial f}{\partial x} + j\frac{\partial f}{\partial y} + k\frac{\partial f}{\partial z},$$

$$h^2\frac{\partial^2 f}{\partial x^2} + j^2\frac{\partial^2 f}{\partial y^2} + k^2\frac{\partial^2 f}{\partial z^2} + 2jk\frac{\partial^2 f}{\partial y \partial z} + 2kh\frac{\partial^2 f}{\partial z \partial x} + 2hj\frac{\partial^2 f}{\partial x \partial y}.$$

In the term

$$\left(h\frac{\partial}{\partial x}+j\frac{\partial}{\partial y}+k\frac{\partial}{\partial z}\right)^{n}f(x+\theta h,y+\theta j,z+\theta k),$$

it is understood that, AFTER the n operations, the variables x, y, z are replaced by $x+\theta h$, $y+\theta j$, $z+\theta k$.

Suppose that x, y, z have definitely assigned values. Form the function $F(t)$ of the variable t by means of the relation

$$F(t)\equiv f(x+th,y+tj,z+tk).$$

If, for convenience, we write

$$u=x+th,\quad v=y+tj,\quad w=z+tk,$$

then (x, y, z being temporarily constant)

$$\frac{dF}{dt}=\frac{\partial f}{\partial u}\frac{du}{dt}+\frac{\partial f}{\partial v}\frac{dv}{dt}+\frac{\partial f}{\partial w}\frac{dw}{dt}$$

$$=h\frac{\partial f}{\partial u}+j\frac{\partial f}{\partial v}+k\frac{\partial f}{\partial w}.$$

(Compare also p. 48.) Thus the operation

$$\frac{d}{dt}$$

on the function $F(t)$ is equivalent to the operation

$$h\frac{\partial}{\partial u}+j\frac{\partial}{\partial v}+k\frac{\partial}{\partial w}$$

on the function $f(u,v,w)$. Applying the same rule to $\dfrac{dF}{dt}$, which is a function of t, and to its alternative expression $h\dfrac{\partial f}{\partial u}+j\dfrac{\partial f}{\partial v}+k\dfrac{\partial f}{\partial w}$, which is a function of u, v, w, we obtain the equivalence

$$\frac{d^{2}}{dt^{2}}\equiv\left(h\frac{\partial}{\partial u}+j\frac{\partial}{\partial v}+k\frac{\partial}{\partial w}\right)^{2},$$

and so on.

Now, by Maclaurin's theorem (Vol. II, p. 49), θ exists ($0<\theta<1$) such that

$$F(t)=F(0)+tF'(0)+\ldots+\frac{t^{n-1}}{(n-1)!}F^{(n-1)}(0)+\frac{t^{n}}{n!}F^{(n)}(\theta t),$$

and so, using the relations just obtained,

$$f(u, v, w)$$

$$= f(u, v, w)_0$$

$$+ t\left(h\frac{\partial}{\partial u} + j\frac{\partial}{\partial v} + k\frac{\partial}{\partial w}\right) f(u, v, w)_0$$

$$\cdots\cdots\cdots\cdots\cdots\cdots\cdots\cdots\cdots$$

$$+ \frac{t^{n-1}}{(n-1)!}\left(h\frac{\partial}{\partial u} + j\frac{\partial}{\partial v} + k\frac{\partial}{\partial w}\right)^{n-1} f(u, v, w)_0$$

$$+ \frac{t^n}{n!}\left(h\frac{\partial}{\partial u} + j\frac{\partial}{\partial v} + k\frac{\partial}{\partial w}\right)^n f(u, v, w)_{\theta t},$$

where, on the right-hand side, the suffix 0 indicates replacement of t by zero in f and its differential coefficients, while the suffix θt indicates replacement by θt, AFTER DIFFERENTIATION in all cases; thus

$$\left(h\frac{\partial}{\partial u} + j\frac{\partial}{\partial v} + k\frac{\partial}{\partial w}\right) f(u, v, w)_0$$

means $\qquad h\dfrac{\partial}{\partial x}f(x, y, z) + j\dfrac{\partial}{\partial y}f(x, y, z) + k\dfrac{\partial}{\partial z}f(x, y, z),$

and $\qquad \left(h\dfrac{\partial}{\partial u} + j\dfrac{\partial}{\partial v} + k\dfrac{\partial}{\partial w}\right)^n f(u, v, w)_{\theta t}$

means $\qquad \left(h\dfrac{\partial}{\partial x} + j\dfrac{\partial}{\partial y} + k\dfrac{\partial}{\partial z}\right)^n f(x + h\theta t, y + j\theta t, z + k\theta t),$

the values $x + h\theta t,\ y + j\theta t,\ z + k\theta t$ being inserted in the partial differential coefficients AFTER differentiation by the current variables x, y, z.

Putting $t = 1$, we obtain the theorem.

ILLUSTRATION 19. *Taylor's theorem for the function xyz.* If

$$f(x, y, z) \equiv xyz,$$

then $\qquad \left(h\dfrac{\partial}{\partial x} + j\dfrac{\partial}{\partial y} + k\dfrac{\partial}{\partial z}\right) f(x, y, z) = hyz + jzx + kxy$

and $\qquad \left(h\dfrac{\partial}{\partial x} + j\dfrac{\partial}{\partial y} + k\dfrac{\partial}{\partial z}\right)^2 f(x, y, z) = 2jkx + 2khy + 2hjz.$

Hence

(i) a number θ exists such that

$(x+h)(y+j)(z+k)$

$= xyz$

$\quad + hyz+jzx+kxy$

$\quad\quad + \tfrac{1}{2}\{2jk(x+\theta h)+2kh(y+\theta j)+2hj(z+\theta k)\},$

so that, on expanding and cancelling like terms from each side,

$$hjk = 3hjk\theta,$$

and $\theta = \tfrac{1}{3};$

(ii) a number ϕ exists such that

$(x+h)(y+j)(z+k)$

$= xyz + h(y+\phi j)(z+\phi k)+j(z+\phi k)(x+\phi h)+k(x+\phi h)(y+\phi j),$

so that, on expanding and cancelling like terms from each side,

$$jkx+khy+hjz+hjk = 2\phi(jkx+khy+hjz)+3\phi^2 hjk.$$

Taking the general case, in which h, j, k are not zero, divide by hjk; then

$$3\phi^2 + (2\phi-1)\left(\frac{x}{h}+\frac{y}{j}+\frac{z}{k}\right) - 1 = 0.$$

It is an interesting exercise in elementary algebra, which we leave to the reader, to prove that, if the expression

$$\frac{x}{h}+\frac{y}{j}+\frac{z}{k}$$

does not lie between $-1, -2$, then there is precisely one value of ϕ between $0, 1$; but that, if the expression does lie between $-1, -2$, then there are two such values of ϕ. In either case, the theorem is verified.

REVISION EXAMPLES X

'Scholarship' Level

1. Prove that, if z is a function of u, v, where

$$u = xy, \quad v = x+y$$

(so that z may also be regarded as a function of x, y), then

$$\frac{\partial^2 z}{\partial x^2} = y^2\frac{\partial^2 z}{\partial u^2} + 2y\frac{\partial^2 z}{\partial u\,\partial v} + \frac{\partial^2 z}{\partial v^2}.$$

2. If the independent variables x, y in the function $f(x, y)$ are changed to ξ, η, where

$$\xi = x+y, \quad \eta = x-y,$$

so that the function $f(x,y)$ becomes $g(\xi, \eta)$, prove that

$$\frac{\partial^2 f}{\partial x^2} - \frac{\partial^2 f}{\partial y^2} = 4\frac{\partial^2 g}{\partial \xi\, \partial \eta}.$$

Deduce, or prove otherwise, that, if the independent variables x, y are changed to u, v, where

$$x+y = \log(u+v), \quad x-y = \log(u-v),$$

so that the function $f(x, y)$ becomes $h(u, v)$, then

$$\frac{\partial^2 f}{\partial x^2} - \frac{\partial^2 f}{\partial y^2} = (u^2 - v^2)\left(\frac{\partial^2 h}{\partial u^2} - \frac{\partial^2 h}{\partial v^2}\right).$$

3. (i) Prove that, in general,

$$\left(x\frac{\partial}{\partial x} + y\frac{\partial}{\partial y}\right)\left(x\frac{\partial}{\partial y} - y\frac{\partial}{\partial x}\right) f(x, y)$$

is *not* equal to $\qquad (x^2 - y^2)\dfrac{\partial^2 f}{\partial x\, \partial y} + xy\left(\dfrac{\partial^2 f}{\partial y^2} - \dfrac{\partial^2 f}{\partial x^2}\right).$

(ii) Prove that, if f is a function of u and

$$u = (x^2 + y^2)\tan^{-1}\frac{y}{x},$$

then $\qquad\qquad x\dfrac{\partial f}{\partial y} - y\dfrac{\partial f}{\partial x} = (x^2 + y^2)\dfrac{df}{du}$

and $\qquad \left(x\dfrac{\partial}{\partial x} + y\dfrac{\partial}{\partial y}\right)\left(x\dfrac{\partial f}{\partial y} - y\dfrac{\partial f}{\partial x}\right) = 2(x^2 + y^2)\left(\dfrac{df}{du} + u\dfrac{d^2 f}{du^2}\right).$

4. If $\qquad\qquad\qquad x = uv, \quad y = u+v,$

prove that $\qquad\qquad \dfrac{\partial x}{\partial v}\dfrac{\partial v}{\partial x} = \dfrac{u}{u-v} = \dfrac{\partial y}{\partial u}\dfrac{\partial u}{\partial y}.$

If z is a function of x, y which satisfies the equation $\dfrac{\partial^2 z}{\partial x\, \partial y} = 0$, prove that

$$(u-v)\left(u\frac{\partial^2 z}{\partial u^2} + v\frac{\partial^2 z}{\partial v^2}\right) - (u^2 - v^2)\frac{\partial^2 z}{\partial u\, \partial v} - (u+v)\left(\frac{\partial z}{\partial u} - \frac{\partial z}{\partial v}\right) = 0.$$

[A permissible method of proving the second part of the question is to assume that z is of the form $f(x) + g(y)$.]

5. If
$$\xi = xy, \quad \eta = x + y,$$

find $\dfrac{\partial x}{\partial \xi}, \dfrac{\partial y}{\partial \xi}$ when η is constant, and $\dfrac{\partial x}{\partial \eta}, \dfrac{\partial y}{\partial \eta}$ when ξ is constant.

Prove that
$$\frac{\partial^2}{\partial \eta\, \partial \xi}(e^x) = \left[\frac{x+y}{(x-y)^3} - \frac{x}{(x-y)^2}\right]e^x.$$

6. If
$$u = e^x(x \cos y - y \sin y),$$

prove that
$$\frac{\partial^2 u}{\partial x^2} + \frac{\partial^2 u}{\partial y^2} = 0,$$

$$e^{-u}\left(\frac{\partial^2 e^u}{\partial x^2} + \frac{\partial^2 e^u}{\partial y^2}\right) = e^{2x}\{(x+1)^2 + y^2\}.$$

If $\quad u = e^x(x \cos y - y \sin y), \quad v = e^x(x \sin y + y \cos y),$

prove that
$$\frac{\partial x}{\partial u} = \frac{e^{-x}\{(x+1)\cos y - y \sin y\}}{(x+1)^2 + y^2},$$

where x is considered as a function of the independent variables u, v.

7. Variables x, y are defined in terms of u, v by the equations
$$x = c \cosh u \cos v, \quad y = c \sinh u \sin v.$$

Write down the values of
$$\frac{\partial x}{\partial u}, \quad \frac{\partial y}{\partial u}, \quad \frac{\partial x}{\partial v}, \quad \frac{\partial y}{\partial v},$$

and prove that
$$\frac{\partial}{\partial u}(x+y)^2 + \frac{\partial}{\partial v}(x+y)^2 = 4c(x+y)\sinh u \cos v,$$

$$\frac{\partial^2}{\partial u^2}(x+y)^2 + \frac{\partial^2}{\partial v^2}(x+y)^2 = 4c^2(\cosh^2 u - \cos^2 v).$$

8. The function $u(x, y)$ is defined by the formula
$$u = e^x(x \cos y - y \sin y).$$

Find a function $v(x, y)$ satisfying the conditions
$$\begin{cases} \dfrac{\partial v}{\partial x} = -\dfrac{\partial u}{\partial y}, \quad \dfrac{\partial v}{\partial y} = \dfrac{\partial u}{\partial x}, \\ v(0, 0) = 0. \end{cases}$$

Show that $u + iv$ is expressible in the form $F(z)$, with $z = x + iy$, and find the function F.

9. (i) Prove that the equation

$$\frac{\partial^2 f}{\partial x^2} + \frac{\partial^2 f}{\partial y^2} + \frac{\partial^2 f}{\partial z^2} + \frac{2f}{x^2+y^2+z^2} = 0$$

is satisfied by $f = \dfrac{x+y+z}{x^2+y^2+z^2}.$

(ii) Given that

$$x = e^{u^2-v^2}\cos 2uv, \quad y = e^{u^2-v^2}\sin 2uv,$$

prove that $\dfrac{\partial x}{\partial u} = \dfrac{\partial y}{\partial v}, \quad \dfrac{\partial y}{\partial u} = -\dfrac{\partial x}{\partial v}.$

Prove that, if f is a function of x, y, then

$$\frac{\partial f}{\partial u} = 2(ux-vy)\frac{\partial f}{\partial x} + 2(uy+vx)\frac{\partial f}{\partial y},$$

$$\frac{\partial f}{\partial v} = -2(uy+vx)\frac{\partial f}{\partial x} + 2(ux-vy)\frac{\partial f}{\partial y}.$$

Hence prove that $\dfrac{\left(\dfrac{\partial f}{\partial u}\right)^2 + \left(\dfrac{\partial f}{\partial v}\right)^2}{\left(\dfrac{\partial f}{\partial x}\right)^2 + \left(\dfrac{\partial f}{\partial y}\right)^2}$

is the same for all functions f.

10. Given that

$$x = c\cosh u \cos v, \quad y = c\sinh u \sin v,$$

where c is constant, find expressions for

$$\frac{\partial x}{\partial u}, \quad \frac{\partial y}{\partial v}.$$

Given that f is a function of u, v, prove that

$$\frac{\partial f}{\partial u} + i\frac{\partial f}{\partial v} = c\sinh(u-iv)\left(\frac{\partial f}{\partial x} + i\frac{\partial f}{\partial y}\right),$$

$$\frac{\partial^2 f}{\partial u^2} + \frac{\partial^2 f}{\partial v^2} = c^2(\cosh^2 u - \cos^2 v)\left(\frac{\partial^2 f}{\partial x^2} + \frac{\partial^2 f}{\partial y^2}\right).$$

11. Given that $x = \dfrac{u}{u^2+v^2}, \quad y = \dfrac{v}{u^2+v^2},$

and that f is a function of x, y, show that

$$(x^2+y^2)\left(\frac{\partial^2 f}{\partial x^2} + \frac{\partial^2 f}{\partial y^2}\right) = (u^2+v^2)\left(\frac{\partial^2 f}{\partial u^2} + \frac{\partial^2 f}{\partial v^2}\right).$$

12. (i) Given that
$$u = x^2 + y^2, \quad v = x^3 + y^3,$$
prove that, if x is considered as a function of u, v,
$$\frac{\partial x}{\partial u} = -\frac{y}{2x(x-y)}, \quad \frac{\partial x}{\partial v} = \frac{1}{3x(x-y)}.$$

Prove that
$$\begin{vmatrix} \dfrac{\partial x}{\partial u} & \dfrac{\partial x}{\partial v} \\[2ex] \dfrac{\partial y}{\partial u} & \dfrac{\partial y}{\partial v} \end{vmatrix} = -\frac{1}{6xy(x-y)}.$$

(ii) The variables x, y, θ are connected by the implicit relation
$$x^2 + y^2 - 2xy\cos\theta = \sin^2\theta.$$
Prove that $\theta = \pm(\sin^{-1}x - \sin^{-1}y)$, and deduce that
$$\frac{\partial^2\theta}{\partial x\,\partial y} = 0.$$

13. If
$$\frac{\sin\theta}{x} = \frac{\sinh\phi}{y} = \cos\theta + \cosh\phi,$$
prove that
$$\frac{\partial x}{\partial\theta} = \frac{\partial y}{\partial\phi}, \quad \frac{\partial x}{\partial\phi} = -\frac{\partial y}{\partial\theta}.$$

The function $u(x,y)$ transforms by means of the above relations into $v(\theta,\phi)$. Prove that
$$\left(\frac{\partial u}{\partial x}\right)^2 + \left(\frac{\partial u}{\partial y}\right)^2 = \frac{\left(\dfrac{\partial v}{\partial\theta}\right)^2 + \left(\dfrac{\partial v}{\partial\phi}\right)^2}{(\cos\theta + \cosh\phi)^2}.$$

14. The identical relation
$$f(u^2 - x^2, u^2 - y^2, u^2 - z^2) = 0$$
defines u as a function of x, y, z. Prove that
$$\frac{1}{x}\frac{\partial u}{\partial x} + \frac{1}{y}\frac{\partial u}{\partial y} + \frac{1}{z}\frac{\partial u}{\partial z} = \frac{1}{u}.$$

15. Prove that, if
$$x = r\sin\theta\cos\phi, \quad y = r\sin\theta\sin\phi, \quad z = r\cos\theta,$$
and if r is a given function of θ, ϕ, so that z may be regarded as a function of x, y, then
$$\left(\frac{\partial r}{\partial\theta}\sin^2\theta + r\sin\theta\cos\theta\right)\frac{\partial z}{\partial x} + \frac{\partial r}{\partial\phi}\sin\phi + r\sin^2\theta\cos\phi$$
$$-\frac{\partial r}{\partial\theta}\sin\theta\cos\theta\cos\phi = 0.$$

16. Prove that, if

$$x = u+v+w, \quad y = vw+wu+uv, \quad z = uvw,$$

and if z is a function of x, y so that w is a function of u, v, then

$$\frac{\dfrac{\partial w}{\partial u}}{vw-p-q(v+w)} = \frac{\dfrac{\partial w}{\partial v}}{wu-p-q(w+u)} = \frac{-1}{uv-p-q(u+v)},$$

where

$$p = \frac{\partial z}{\partial x}, \quad q = \frac{\partial z}{\partial y}.$$

17. Show that, if x, y are functions of ξ, η, and if

$$a \equiv \left(\frac{\partial x}{\partial \xi}\right)^2 + \left(\frac{\partial y}{\partial \xi}\right)^2, \quad h \equiv \frac{\partial x}{\partial \xi}\frac{\partial x}{\partial \eta} + \frac{\partial y}{\partial \xi}\frac{\partial y}{\partial \eta}, \quad b \equiv \left(\frac{\partial x}{\partial \eta}\right)^2 + \left(\frac{\partial y}{\partial \eta}\right)^2,$$

$$H \equiv \frac{\partial x}{\partial \xi}\frac{\partial y}{\partial \eta} - \frac{\partial x}{\partial \eta}\frac{\partial y}{\partial \xi},$$

then, for any differentiable function $U(x, y)$,

$$H\frac{\partial U}{\partial x} = \frac{\partial U}{\partial \xi}\frac{\partial y}{\partial \eta} - \frac{\partial U}{\partial \eta}\frac{\partial y}{\partial \xi},$$

$$H\frac{\partial U}{\partial y} = -\frac{\partial U}{\partial \xi}\frac{\partial x}{\partial \eta} + \frac{\partial U}{\partial \eta}\frac{\partial x}{\partial \xi}.$$

Prove also that

$$\frac{\partial^2 U}{\partial x^2} + \frac{\partial^2 U}{\partial y^2} = \frac{1}{H}\frac{\partial}{\partial \xi}\left\{\frac{b\dfrac{\partial U}{\partial \xi} - h\dfrac{\partial U}{\partial \eta}}{H}\right\} - \frac{1}{H}\frac{\partial}{\partial \eta}\left\{\frac{h\dfrac{\partial U}{\partial \xi} - a\dfrac{\partial U}{\partial \eta}}{H}\right\}.$$

18. A function $f(x, y)$, when expressed in terms of the new variables u, v defined by the equations

$$x = \tfrac{1}{2}(u+v), \quad y^2 = uv,$$

becomes $g(u, v)$. Prove that

$$\frac{\partial^2 g}{\partial u\,\partial v} = \frac{1}{4}\left(\frac{\partial^2 f}{\partial x^2} + 2\frac{x}{y}\frac{\partial^2 f}{\partial x\,\partial y} + \frac{\partial^2 f}{\partial y^2} + \frac{1}{y}\frac{\partial f}{\partial y}\right).$$

19. Prove that, if $f(x, y)$ is a function of $x^2 + y^2$ only, it satisfies the identical relation

$$y\frac{\partial f}{\partial x} = x\frac{\partial f}{\partial y}.$$

By changing to polar coordinates, or otherwise, prove conversely that, if $f(x, y)$ satisfies this relation identically, then it must be a function of $x^2 + y^2$ only.

20. Prove that, if x, y, z are functions of two variables u, v given by the relations

$$x = f(u, v), \quad y = g(u, v), \quad z = h(u, v),$$

then (suffixes denoting partial differentiation)

$$(f_u g_v - f_v g_u) \frac{\partial z}{\partial x} = h_u g_v - h_v g_u,$$

$$(f_u g_v - f_v g_u)^3 \frac{\partial^2 z}{\partial x^2} = \begin{vmatrix} f_u & f_v & f_{uu}g_v^2 - 2f_{uv}g_u g_v + f_{vv}g_u^2 \\ g_u & g_v & g_{uu}g_v^2 - 2g_{uv}g_u g_v + g_{vv}g_u^2 \\ h_u & h_v & h_{uu}g_v^2 - 2h_{uv}g_u g_v + h_{vv}g_u^2 \end{vmatrix}.$$

21. The function y of two variables θ, x is defined implicitly by the equation $$y = \theta + x\phi(y).$$

If u is a function of y and $F(u)$ a function of u, prove that

$$\frac{\partial}{\partial \theta}\left\{ F(u) \frac{\partial u}{\partial x} \right\} = \frac{\partial}{\partial x}\left\{ F(u) \frac{\partial u}{\partial \theta} \right\},$$

and that $\dfrac{\partial u}{\partial x} = \phi(y)\dfrac{\partial u}{\partial \theta}, \quad \dfrac{\partial^2 u}{\partial x^2} = \dfrac{\partial}{\partial \theta}\left[\{\phi(y)\}^2 \dfrac{\partial u}{\partial \theta} \right],$

and, generally, that

$$\frac{\partial^n u}{\partial x^n} = \frac{\partial^{n-1}}{\partial \theta^{n-1}}\left[\{\phi(y)\}^n \frac{\partial u}{\partial \theta} \right].$$

22. Prove that, if
$$x = r\cos\theta, \quad y = r\sin\theta,$$

and if ϕ is any function of r, θ, then

$$\frac{\partial \phi}{\partial x} = \frac{\partial \phi}{\partial r}\cos\theta - \frac{\partial \phi}{\partial \theta}\frac{\sin\theta}{r},$$

and obtain a corresponding expression for $\dfrac{\partial \phi}{\partial y}$.

Prove that, if $\phi = r^{-n}\sin n\theta,$

then $\dfrac{\partial^2 \phi}{\partial x^2} + \dfrac{\partial^2 \phi}{\partial y^2} = 0.$

23. Prove that, if
$$U = f(x/y), \quad U_n = r^n U,$$

where $r^2 = x^2 + y^2$, then $x\dfrac{\partial U}{\partial x} + y\dfrac{\partial U}{\partial y} = 0,$

$$\frac{\partial^2 U_n}{\partial x^2} + \frac{\partial^2 U_n}{\partial y^2} = r^n\left(\frac{\partial^2 U}{\partial x^2} + \frac{\partial^2 U}{\partial y^2} + \frac{n^2}{r^2} U \right).$$

Hence, or otherwise, prove that, if the equation

$$\frac{\partial^2 u}{\partial x^2} + \frac{\partial^2 u}{\partial y^2} = 0$$

is satisfied by $u = V(x, y)$, where $V(x, y)$ is a homogeneous function of degree n, then the equation is satisfied also by $u = r^{-2n} V(x, y)$.

24. Four variables u_1, u_2, u_3, u_4 are such that any three are independent and each may be expressed as a function of the other three. If u_{12} denotes $\dfrac{\partial u_1}{\partial u_2}$, the rate of change of u_1 with respect to u_2 when u_3, u_4 are kept fixed, and if u_{ij} ($i \neq j$) has a similar meaning, show that

$$u_{23} u_{31} u_{12} = -1, \quad u_{12} u_{34} = u_{14} u_{32}.$$

25. The variables x, y, z satisfy the two equations

$$f(x, y, z) = 0, \quad g(x, y, z) = 0.$$

By eliminating z between these equations, it is possible to obtain a relation connecting x, y which defines y as a function of x. Show that

$$\frac{dy}{dx} = -\frac{f_x g_z - f_z g_x}{f_y g_z - f_z g_y}.$$

26. Given that x, y are functions of u, v defined by

$$f(x, y, u, v) = 0, \quad g(x, y, u, v) = 0,$$

find $\dfrac{\partial x}{\partial u}$ in terms of partial derivatives of f, g with respect to x, y, u, v.

If
$$x^2 + y^2 - 25uv = 0,$$

$$ux + vy - 1 = 0,$$

prove that $\dfrac{\partial x}{\partial u} = \pm \frac{1}{14}$ when $u = v = 1$, and give the reason for the ambiguity in sign.

27. If $f(x, y)$ is a function of x, y, where x, y are functions of t defined by the relations

$$u(x, t) = 0, \quad v(y, t) = 0,$$

and if $f(x, y)$, when expressed as a function of t, takes the form $g(t)$, prove that

$$\frac{dg}{dt} = -\left(\frac{\partial f}{\partial x}\frac{\partial v}{\partial y}\frac{\partial u}{\partial t} + \frac{\partial f}{\partial y}\frac{\partial u}{\partial x}\frac{\partial v}{\partial t}\right) \bigg/ \frac{\partial u}{\partial x}\frac{\partial v}{\partial y}.$$

28. (i) If
$$u = xyz,$$

where x, y, z are connected by the relations

$$yz + zx + xy = a, \quad x + y + z = b \quad (a, b \text{ constants}),$$

prove that
$$\frac{du}{dx} = (x - y)(x - z).$$

(ii) If ξ, η are functions of x, y such that

$$\xi = e^x \cos y, \quad \eta = e^x \sin y,$$

and if x, y are functions of r, θ such that

$$x = e^r \cos \theta, \quad y = e^r \sin \theta,$$

where r is a function of θ, prove that

$$\frac{d\xi}{d\eta} = \frac{\dfrac{dr}{d\theta} - \tan(y + \theta)}{1 + \dfrac{dr}{d\theta} \tan(y + \theta)}.$$

29. The equation
$$z = F(x, y)$$

is obtained by eliminating u between the equations

$$y = f(u, x), \quad z = g(u, x).$$

Prove that
$$\frac{\partial f}{\partial u} \frac{\partial F}{\partial x} = \frac{\partial f}{\partial u} \frac{\partial g}{\partial x} - \frac{\partial f}{\partial x} \frac{\partial g}{\partial u},$$

$$\frac{\partial f}{\partial u} \frac{\partial F}{\partial y} = \frac{\partial g}{\partial u}.$$

30. Prove that, if u is a function of x, y which is transformed into a function of r, θ by the relations

$$x = r^{\frac{1}{2}} \cos \tfrac{1}{2}\theta, \quad y = r^{\frac{1}{2}} \sin \tfrac{1}{2}\theta,$$

then
(i) $2r \dfrac{\partial u}{\partial r} = x \dfrac{\partial u}{\partial x} + y \dfrac{\partial u}{\partial y},$

(ii) $2 \dfrac{\partial u}{\partial \theta} = x \dfrac{\partial u}{\partial y} - y \dfrac{\partial u}{\partial x},$

(iii) $4\left(r \dfrac{\partial^2 u}{\partial r^2} + \dfrac{\partial u}{\partial r} + \dfrac{1}{r} \dfrac{\partial^2 u}{\partial \theta^2} \right) = \dfrac{\partial^2 u}{\partial x^2} + \dfrac{\partial^2 u}{\partial y^2}.$

31. Given that $x = r \cos \theta, \quad y = r \sin \theta,$

find $\dfrac{\partial \theta}{\partial x}, \quad \dfrac{\partial \theta}{\partial y}, \quad \dfrac{\partial^2 \theta}{\partial x \, \partial y},$

and prove that $\dfrac{\partial^{2n} \theta}{\partial x^n \, \partial y^n} = \dfrac{(2n-1)!}{r^{2n}} \sin (2n\theta - \tfrac{1}{2}n\pi).$

32. If x, y are connected with r, θ by the relations

$$x \cos \theta - y \sin \theta = x \sin \theta + y \cos \theta = r,$$

find $\dfrac{\partial \theta}{\partial x}, \quad \dfrac{\partial r}{\partial x}, \quad \dfrac{\partial \theta}{\partial y}, \quad \dfrac{\partial r}{\partial y}$

in terms of r, θ.

Show that, if V is a function of x, y,

$$\frac{\partial^2 V}{\partial x^2} + \frac{\partial^2 V}{\partial y^2} = \frac{1}{2} \left(\frac{\partial^2 V}{\partial r^2} + \frac{1}{r} \frac{\partial V}{\partial r} + \frac{1}{r^2} \frac{\partial^2 V}{\partial \theta^2} \right).$$

33. The variables x, y are changed to r, θ, where

$$x = r \sec \theta, \quad y = r \tan \theta.$$

Show that

$$\frac{\partial^2 u}{\partial x^2} - \frac{\partial^2 u}{\partial y^2} = \frac{\partial^2 u}{\partial r^2} + \frac{1}{r} \frac{\partial u}{\partial r} - \frac{\cos^2 \theta}{r^2} \frac{\partial^2 u}{\partial \theta^2} + \frac{\sin \theta \cos \theta}{r^2} \frac{\partial u}{\partial \theta}.$$

34. Given that u, v are functions of the two variables x, y, express $\left(\dfrac{\partial^2}{\partial x^2} + \dfrac{\partial^2}{\partial y^2} \right) (uv)$ in terms of the partial differential coefficients of u and v.

If u is a function, with third-order differential coefficients, satisfying the relation $\dfrac{\partial^2 u}{\partial x^2} + \dfrac{\partial^2 u}{\partial y^2} = 0$, and if

$$w = (x^2 + y^2 + ax + by + c)\, u,$$

show that $\left(\dfrac{\partial^2}{\partial x^2} + \dfrac{\partial^2}{\partial y^2} \right)^2 w = 0.$

35. Four variables x, y, r, θ are connected by the two relations

$$x = r \cos \theta, \quad y = r \sin \theta.$$

A symbol like $\left(\dfrac{\partial r}{\partial x} \right)_y$ indicates that r is expressed as a function of x, y and that y is kept constant during the differentiation of r with respect to x. Verify the relations

(i) $\left(\dfrac{\partial x}{\partial r} \right)_\theta \left(\dfrac{\partial r}{\partial x} \right)_\theta = 1,$ 　　　(ii) $\left(\dfrac{\partial x}{\partial r} \right)_\theta = \left(\dfrac{\partial r}{\partial x} \right)_y,$

(iii) $\left(\dfrac{\partial y}{\partial x} \right)_\theta = - \left(\dfrac{\partial x}{\partial y} \right)_r,$ 　　　(iv) $\dfrac{1}{r} \left(\dfrac{\partial x}{\partial \theta} \right)_r = r \left(\dfrac{\partial \theta}{\partial x} \right)_y.$

Illustrate (iii), (iv) geometrically, taking (x, y), (r, θ) as Cartesian and polar coordinates of a point in a plane.

36. If Δ is the area of a triangle ABC, calculate $\dfrac{\partial \Delta}{\partial a}$ when the independent variables are

(i) a, b, C;   (ii) a, B, C;   (iii) a, b, c.

Illustrate your results geometrically.

37. If the circumradius R, and the area Δ, of a triangle ABC are regarded as functions of b, c, A, prove that

$$\frac{\partial R}{\partial A} = 4R \sin A \frac{\partial R}{\partial b} \frac{\partial R}{\partial c},$$

$$\frac{\partial R}{\partial b} \frac{\partial \Delta}{\partial c} + \frac{\partial R}{\partial c} \frac{\partial \Delta}{\partial b} = \tfrac{1}{2} R \sin A.$$

38. If (x, y) are the Cartesian coordinates of a point and (r, θ) its polar coordinates, define what you mean by $\dfrac{\partial x}{\partial r}, \dfrac{\partial r}{\partial x}$.

Find their values, and illustrate by a diagram.

Prove that
$$\frac{\partial^2 r}{\partial x^2} \frac{\partial^2 r}{\partial y^2} = \left(\frac{\partial^2 r}{\partial x \partial y} \right)^2.$$

39. If $u = f(r)$, where

$$x = r \cos \theta, \quad y = r \sin \theta,$$

prove that
$$\frac{\partial^2 u}{\partial x^2} + \frac{\partial^2 u}{\partial y^2} = \frac{d^2 f}{dr^2} + \frac{1}{r} \frac{df}{dr}.$$

40. If
$$z = x^r f(y/x),$$

prove that
$$x \frac{\partial z}{\partial x} + y \frac{\partial z}{\partial y} = rz,$$

$$x^2 \frac{\partial^2 z}{\partial x^2} + 2xy \frac{\partial^2 z}{\partial x \partial y} + y^2 \frac{\partial^2 z}{\partial y^2} = r(r-1) z.$$

41. If
$$u = f(ax^2 + 2hxy + by^2),$$
$$v = g(ax^2 + 2hxy + by^2),$$

prove that
$$\frac{\partial}{\partial y} \left(u \frac{\partial v}{\partial x} \right) = \frac{\partial}{\partial x} \left(u \frac{\partial v}{\partial y} \right).$$

42. A function f of two independent variables r, θ is transformed into a function g of variables u, s by means of the relations

$$r \cos \theta = 1/u, \quad \tan \theta = s.$$

Prove that $\quad r \dfrac{\partial f}{\partial r} = -u \dfrac{\partial g}{\partial u}, \quad \dfrac{\partial f}{\partial \theta} = us \dfrac{\partial g}{\partial u} + (1 + s^2) \dfrac{\partial g}{\partial s}.$

Prove also that, if f satisfies the equation

$$\cos \theta \frac{\partial^2 f}{\partial r \, \partial \theta} + r \sin \theta \frac{\partial^2 f}{\partial r^2} = 0,$$

then $\qquad\qquad \dfrac{\partial g}{\partial u} = (1 + s^2)^{\frac{1}{2}} \phi(u),$

where ϕ is some function of u.

43. If x, y are the coordinates of any point and

$$x = r \cos \theta, \quad y = r \sin \theta,$$

find the values of $\quad \left(\dfrac{\partial x}{\partial \theta} \right)_{r \, \text{const.}}, \quad \left(\dfrac{\partial \theta}{\partial x} \right)_{y \, \text{const.}}.$

Verify the second result geometrically.

Prove that $\qquad\qquad \dfrac{\partial^2 \theta}{\partial x^2} + \dfrac{\partial^2 \theta}{\partial y^2} = 0.$

44. The variables x, y, z are connected by the relation

$$x^2 + y^2 + z^2 - 3xyz = 0,$$

and $\qquad\qquad \phi(x, y, z) \equiv x^3 y^2 z.$

Determine the value of $\dfrac{\partial \phi}{\partial x}$ at $(1, 1, 1)$,

(i) when the independent variables are x, y,

(ii) when they are x, z,

and explain geometrically the difference between the meanings of $\dfrac{\partial \phi}{\partial x}$ in the two cases.

45. Each of u, v, w is a function of the three variables x, y, z. Whenever x, y receive small increments which satisfy the equation

$dy - z\,dx = 0$, the corresponding increments du, dv satisfy the equation $dv - w\,du = 0$. Prove that

$$\frac{\partial v}{\partial x} - w\frac{\partial u}{\partial x} + z\left(\frac{\partial v}{\partial y} - w\frac{\partial u}{\partial y}\right) = 0,$$

$$\frac{\partial v}{\partial z} - w\frac{\partial u}{\partial z} = 0.$$

Find the most general functions u, v, w for which u is a function of x only, v of y only, and w of z only.

46. If $\phi(x,y)$ is a differentiable function of x, y, and if

$$x = \frac{\sin u}{\cos u + \cosh v}, \quad y = \frac{f(v)}{\cos u + \cosh v},$$

show that, for a certain form of the function $f(v)$,

$$x\frac{\partial \phi}{\partial x} + y\frac{\partial \phi}{\partial y} = \sin u \cosh v \frac{\partial \phi}{\partial u} + \cos u \sinh v \frac{\partial \phi}{\partial v},$$

and find the form of $f(v)$ in this case.

47. If $u = x/y$, $v = xy$, prove that

$$x^2\frac{\partial^2 f}{\partial x^2} + y^2\frac{\partial^2 f}{\partial y^2} = 2\left(u^2\frac{\partial^2 f}{\partial u^2} + v^2\frac{\partial^2 f}{\partial v^2} + u\frac{\partial f}{\partial u}\right),$$

$$x^2\frac{\partial^2 f}{\partial x^2} - y^2\frac{\partial^2 f}{\partial y^2} = 2\left(2uv\frac{\partial^2 f}{\partial u\,\partial v} - u\frac{\partial f}{\partial u}\right).$$

48. The area Δ of a triangle is found from measurements of the side a and the angles B, C. Prove that the error $\delta\Delta$ in the calculated value of the area due to small errors δa, δB, δC is given approximately by

$$\frac{\delta\Delta}{\Delta} = 2\frac{\delta a}{a} + \frac{c}{a\sin B}\delta B + \frac{b}{a\sin C}\delta C.$$

49. The side b of a triangle ABC is calculated from measurements of the side a and the angles B, C. There may be an error not exceeding h (in either direction) in the measurement of a and an error not exceeding α in either or both of B, C, where h, α are small. Show that, if $B + C < \frac{1}{2}\pi$, the greatest error in b is approximately

$$b\left\{\frac{h}{a} + \alpha\cot B\right\}.$$

Find the greatest possible error in b when $B + C > \frac{1}{2}\pi$, and explain the difference in the two cases.

50. If
$$f(x, y) \equiv axy + bx + cy + d,$$
$$g(x, y) \equiv a'xy + b'x + c'y + d',$$

prove that
$$\frac{\partial}{\partial x}\left\{\frac{f}{g}\right\} = \frac{p(y)}{g^2},$$

where $p(y)$ is a quadratic function of y (not containing x).

Hence or otherwise prove that $f(x,y)/g(x,y)$ is a function of the product XY, where X is an appropriate function of x only and Y a function of y only.

51. A function y of x is defined by the equation $f(x,y) = 0$. Express $\dfrac{dy}{dx}, \dfrac{d^2y}{dx^2}$ in terms of p, q, r, s, t, where

$$p = \frac{\partial f}{\partial x}, \quad q = \frac{\partial f}{\partial y},$$

$$r = \frac{\partial^2 f}{\partial x^2}, \quad s = \frac{\partial^2 f}{\partial x\,\partial y} = \frac{\partial^2 f}{\partial y\,\partial x}, \quad t = \frac{\partial^2 f}{\partial y^2}.$$

If the same equation is regarded as defining x as a function of y, prove that
$$p^3 \frac{d^2x}{dy^2} = q^3 \frac{d^2y}{dx^2}.$$

52. Show that, if $x = \rho \cos \phi$, $y = \rho \sin \phi$, then
$$\frac{\partial^2 V}{\partial x^2} + \frac{\partial^2 V}{\partial y^2} = \frac{\partial^2 V}{\partial \rho^2} + \frac{1}{\rho}\frac{\partial V}{\partial \rho} + \frac{1}{\rho^2}\frac{\partial^2 V}{\partial \phi^2}.$$

Hence show that
$$\tan^{-1}\frac{y}{x}, \quad z\tan^{-1}\frac{y}{x}, \quad (p^2 + z^2)^{-\frac{1}{2}}\tan^{-1}\frac{y}{x}$$

are all solutions of the equation
$$\frac{\partial^2 V}{\partial x^2} + \frac{\partial^2 V}{\partial y^2} + \frac{\partial^2 V}{\partial z^2} = 0.$$

CHAPTER XV

MAXIMA AND MINIMA

1. The general conditions. Suppose that

$$u \equiv f(x, y, z)$$

is a function of the three variables x, y, z (not necessarily independent). The function u is said to have a MAXIMUM at a point P,
for which $x = a$, $y = b$, $z = c$, if its value at P exceeds that at
neighbouring points, and a MINIMUM if the value is less. As for
functions of one variable, maxima and minima are *local* properties.

The argument for several variables is naturally more complicated
than for one only, and the treatment which follows is of necessity
more sketchy.

We begin by finding *necessary conditions*, assuming first that the
three variables x, y, z are independent.

Suppose that the function

$$u \equiv f(x, y, z)$$

has a maximum value when $x = a$, $y = b$, $z = c$. Its value is then
greater than that at any near point; in particular, at all those near
points for which $y = b$, $z = c$. Thus the function

$$f(x, b, c)$$

of the *single* variable x is greater for $x = a$ than for any other near
value, so that this function has a maximum at $x = a$. Its differential
coefficient with respect to x (while y, z remain constant at b, c
respectively) is therefore zero when $x = a$. Hence

$$\frac{\partial f}{\partial x} = 0$$

when $x = a$, $y = b$, $z = c$. Similarly

$$\frac{\partial f}{\partial y} = 0, \quad \frac{\partial f}{\partial z} = 0$$

for those values.

The presence of a minimum may also be treated in the same way.
Hence *a set of* necessary *conditions for the function*

$$u \equiv f(x, y, z) \quad (x, y, z \ independent)$$

to have a maximum or a minimum value for $x = a$, $y = b$, $z = c$ is that

$$\frac{\partial f}{\partial x} = 0, \quad \frac{\partial f}{\partial y} = 0, \quad \frac{\partial f}{\partial z} = 0$$

for $x = a$, $y = b$, $z = c$.

Since these conditions must hold at each maximum or minimum, they enable us to locate such turning points. But the conditions are *not* sufficient; moreover, they do not, of course, distinguish between maxima and minima. Detailed analysis is beyond the scope of this book, but common sense will often provide the answer for particular problems.

The three conditions may be gathered into a single differential form. For the differential of the function u is

$$du \equiv \frac{\partial f}{\partial x} dx + \frac{\partial f}{\partial y} dy + \frac{\partial f}{\partial z} dz,$$

which vanishes at a maximum or minimum. Hence *the differential $du \equiv \frac{\partial f}{\partial x} dx + \frac{\partial f}{\partial y} dy + \frac{\partial f}{\partial z} dz$ vanishes when u has a maximum or minimum value, for arbitrary values of the differentials dx, dy, dz.*

Expressed in this form, the condition $du = 0$ for a maximum or a minimum may be applied *whether x, y, z are independent or not.* For if they are not independent, the conditions of dependence enable us to reduce the number of variables until those which remain *are* independent, and the condition $du = 0$ then follows.

Note. If x, y, z are not independent, the condition

$$du = 0$$

for a maximum or minimum is still equivalent to

$$\frac{\partial u}{\partial x} dx + \frac{\partial u}{\partial y} dy + \frac{\partial u}{\partial z} dz = 0,$$

but it does NOT follow now that $\frac{\partial u}{\partial x}$, $\frac{\partial u}{\partial y}$, $\frac{\partial u}{\partial z}$ vanish separately. For the treatment to be adopted, see § 3.

ILLUSTRATION 1. *To find the greatest distance of a point on the surface*

$$ax^2 + by^2 + cz^2 + 2fyz + 2gzx + 2hxy = 1$$

(assumed to be an 'ellipsoid') from the plane $z = 0$.

(The surface may be pictured as a somewhat distorted sphere with its centre at the origin.)

The distance of the point (x, y, z) on the surface from the plane $z = 0$ is simply the coordinate z, and, at a point of maximum (or minimum) distance the two partial differential coefficients $\dfrac{\partial z}{\partial x}, \dfrac{\partial z}{\partial y}$ must vanish.

Differentiate the equation of the surface partially with respect to x, and then put $\dfrac{\partial z}{\partial x} = 0$. Thus

$$ax + hy + gz = 0.$$

Similarly,
$$hx + by + fz = 0.$$

These two equations, together with the equation of the surface, determine the points of greatest distance; the z coordinates give the actual distances.

In order to find z, we eliminate x, y as follows:

Multiply the two equations of condition by x, y respectively, and subtract from the equation of the surface; divide the resulting equation by z. Thus

$$gx + fy + cz - \frac{1}{z} = 0.$$

Eliminating $x : y : 1$ determinantally between the three linear equations, we have

$$\begin{vmatrix} a & h & gz \\ h & b & fz \\ g & f & cz - \dfrac{1}{z} \end{vmatrix} = 0,$$

or, after division by z,

$$\begin{vmatrix} a & h & g \\ h & b & f \\ g & f & c \end{vmatrix} + \begin{vmatrix} a & h & 0 \\ h & b & 0 \\ g & f & -\dfrac{1}{z^2} \end{vmatrix} = 0.$$

Expanding,

$$(abc + 2fgh - af^2 - bg^2 - ch^2) - \frac{1}{z^2}(ab - h^2) = 0,$$

so that
$$z = \pm \sqrt{\left(\frac{ab - h^2}{abc + 2fgh - af^2 - bg^2 - ch^2} \right)}.$$

From the shape of the surface, this gives the *greatest* distance, whose value is therefore

$$\sqrt{\left(\frac{ab - h^2}{abc + 2fgh - af^2 - bg^2 - ch^2} \right)}.$$

2.* To distinguish between maxima and minima. Though detailed consideration would carry us much too far, the general picture may be exhibited of a method to distinguish systematically between maxima and minima.

For, say, three independent variables, suppose that the function

$$u \equiv f(x, y, z)$$

has a turning value at the point (a, b, c). It will be a convenience of notation to express the conditions in the form

$$\frac{\partial u}{\partial a} = 0, \quad \frac{\partial u}{\partial b} = 0, \quad \frac{\partial u}{\partial c} = 0.$$

If ξ, η, ζ are small, the value of the function at the near point $(a+\xi, b+\eta, c+\zeta)$ is

$$f(a+\xi, b+\eta, c+\zeta),$$

where, by Taylor's theorem (p. 57),

$$f(a+\xi, b+\eta, c+\zeta) - f(a, b, c)$$

$$= \left(\xi \frac{\partial}{\partial x} + \eta \frac{\partial}{\partial y} + \zeta \frac{\partial}{\partial z} \right) f(x, y, z)$$

$$+ \frac{1}{2!} \left(\xi \frac{\partial}{\partial x} + \eta \frac{\partial}{\partial y} + \zeta \frac{\partial}{\partial z} \right)^2 f(x, y, z)$$

$$+ \dots,$$

the differential coefficients being evaluated at $x = a$, $y = b$, $z = c$. Taking ξ, η, ζ to be so small that powers and products of degree greater than 2 may be neglected, and remembering that

$$\frac{\partial u}{\partial a} = \frac{\partial u}{\partial b} = \frac{\partial u}{\partial c} = 0,$$

we obtain a relation which we may write in the form

$$f(a+\xi, b+\eta, c+\zeta) - f(a, b, c)$$

$$= \frac{1}{2} \left\{ \frac{\partial^2 u}{\partial a^2} \xi^2 + \frac{\partial^2 u}{\partial b^2} \eta^2 + \frac{\partial^2 u}{\partial c^2} \zeta^2 + 2 \frac{\partial^2 u}{\partial b \partial c} \eta\zeta + 2 \frac{\partial^2 u}{\partial c \partial a} \zeta\xi + 2 \frac{\partial^2 u}{\partial a \partial b} \xi\eta \right\},$$

where ξ, η, ζ are variables, depending on the point selected near to (a, b, c), and where the coefficients $\dfrac{\partial^2 u}{\partial a^2}, \dots$ are constants.

---

* This paragraph may be postponed, if desired.

For a MAXIMUM this expression must be negative for *all* sets of values of ξ, η, ζ; for a MINIMUM it must be positive. This condition is both necessary and sufficient. If, however, the expression varies in sign, being positive for some values of ξ, η, ζ and negative for others, the function u has neither maximum nor minimum.

The simpler cases of genuine maxima or minima can be settled by definite algebraic tests:

(i) For convenience, write the expression with the notation

$$E \equiv A\xi^2 + B\eta^2 + C\zeta^2 + 2F\eta\zeta + 2G\zeta\xi + 2H\xi\eta,$$

and form the determinant whose rows are the coefficients of 2ξ, 2η, 2ζ in the partial differential coefficients $\dfrac{\partial E}{\partial \xi}$, $\dfrac{\partial E}{\partial \eta}$, $\dfrac{\partial E}{\partial \zeta}$ respectively, namely,

$$\begin{vmatrix} A & H & G \\ H & B & F \\ G & F & C \end{vmatrix}.$$

Write down also the 'leading diagonal' determinants of 2, 1 rows, namely,

$$\begin{vmatrix} A & H \\ H & B \end{vmatrix}, \quad A.$$

Then it is known that *a necessary and sufficient condition for E to be always positive, for all values of ξ, η, ζ, is that these three determinants should all be positive.*

Consider, for example, the expression

$$3\xi^2 + 2\eta^2 + 2\zeta^2 + 2\eta\zeta - 4\zeta\xi - 4\xi\eta.$$

The determinants are

$$\begin{vmatrix} 3 & -2 & -2 \\ -2 & 2 & 1 \\ -2 & 1 & 2 \end{vmatrix} = 1,$$

$$\begin{vmatrix} 3 & -2 \\ -2 & 2 \end{vmatrix} = 2,$$

and   3.

Since they are all positive, the expression is positive for all ξ, η, ζ.

(ii) The test for E to be always negative is that $-E$ should be always positive.

Thus *the function* $f(x,y)$ *of two independent variables has a* MAXIMUM *at the point* (a,b) *if*

$$\frac{\partial f}{\partial a} = 0, \quad \frac{\partial f}{\partial b} = 0,$$

$$\frac{\partial^2 f}{\partial a^2} < 0,$$

$$\frac{\partial^2 f}{\partial a^2}\frac{\partial^2 f}{\partial b^2} - \left(\frac{\partial^2 f}{\partial a\,\partial b}\right)^2 > 0;$$

and a MINIMUM *if*
$$\frac{\partial f}{\partial a} = 0, \quad \frac{\partial f}{\partial b} = 0,$$

$$\frac{\partial^2 f}{\partial a^2} > 0,$$

$$\frac{\partial^2 f}{\partial a^2}\frac{\partial^2 f}{\partial b^2} - \left(\frac{\partial^2 f}{\partial a\,\partial b}\right)^2 > 0.$$

(Note the direction of the last inequality.)

(iii) The method can be extended, in an obvious way, to any number of variables. For the expression

$$A\xi^2 + B\eta^2 + C\zeta^2 + D\tau^2 + 2F\eta\zeta + 2G\zeta\xi + 2H\xi\eta + 2U\xi\tau + 2V\eta\tau + 2W\zeta\tau,$$

the determinants are

$$\begin{vmatrix} A & H & G & U \\ H & B & F & V \\ G & F & C & W \\ U & V & W & D \end{vmatrix}, \quad \begin{vmatrix} A & H & G \\ H & B & F \\ G & F & C \end{vmatrix}, \quad \begin{vmatrix} A & H \\ H & B \end{vmatrix}, \quad A.$$

3. Lagrange's method of undetermined multipliers. The method now to be explained may be grasped more readily if we begin with an illustration:

ILLUSTRATION 2.* *To find the lengths of the axes of the conic in which the quadric*
$$ax^2 + by^2 + cz^2 + 2fyz + 2gzx + 2hxy = 1$$
meets the plane $\qquad lx + my + nz = 0.$

---

* The reader who has not yet reached the theory of quadrics may simply regard the problem as the determination of the maximum and minimum values of the function $x^2 + y^2 + z^2$, where the variables x, y, z are subject to the two equations quoted in the enunciation.

The squares of the lengths of the axes are the maxima and minima of the function
$$u \equiv x^2 + y^2 + z^2.$$
At such points, $du = 0$, so that
$$x\,dx + y\,dy + z\,dz = 0.$$
Since the point (x, y, z) lies on the quadric, the differentials satisfy the equation
$$(ax + hy + gz)\,dx + (hx + by + fz)\,dy + (gx + fy + cz)\,dz = 0;$$
and, since it lies on the plane,
$$l\,dx + m\,dy + n\,dz = 0.$$
Multiply these three equations among the differentials by $1, \lambda, \mu$ respectively, and add. Then choose λ, μ so that the coefficients of dx, dy both vanish; that is, so that
$$x + \lambda(ax + hy + gz) + \mu l = 0,$$
$$y + \lambda(hx + by + fz) + \mu m = 0.$$
Then $$[z + \lambda(gx + fy + cz) + \mu n]\,dz = 0.$$

Now the *three* variables x, y, z are subject to *two* conditions, namely, the equation of the quadric and the equation of the plane; their three differentials are therefore subject to two linear conditions, so that *one* differential may be given an arbitrarily assigned value. In particular, dz may be assigned arbitrarily—it need not be zero. Hence the coefficient of dz in the equation last written must be zero, so that
$$z + \lambda(gx + fy + cz) + \mu n = 0.$$

To summarize, the *five* equations
$$ax^2 + by^2 + cz^2 + 2fyz + 2gzx + 2hxy = 1,$$
$$lx + my + nz = 0,$$
$$x + \lambda(ax + hy + gz) + \mu l = 0,$$
$$y + \lambda(hx + by + fz) + \mu m = 0,$$
$$z + \lambda(gx + fy + cz) + \mu n = 0$$
serve to determine the *five* unknowns consisting of (i) the three variables x, y, z, evaluated at a turning point, (ii) the two multipliers λ, μ. The sacrifice involved in increasing the number of unknowns is amply repaid in symmetry.

In problems of this type, where the equations of condition are homogeneous polynomials in x, y, z, a good first move is to multiply the equations involving λ, μ by x, y, z respectively, and then to add them. This gives

$$(x^2 + y^2 + z^2)$$
$$+ \lambda(ax^2 + by^2 + cz^2 + 2fyz + 2gzx + 2hxy)$$
$$+ \mu(lx + my + nz) = 0,$$

so that $$u + \lambda(1) + \mu(0) = 0,$$

or $$u + \lambda = 0,$$

or $$\lambda = -u.$$

Hence $$x - u(ax + hy + gz) + \mu l = 0,$$

or $$\left(a - \frac{1}{u}\right)x + hy + gz - \frac{\mu}{u}l = 0.$$

Similarly, $$hx + \left(b - \frac{1}{u}\right)y + fz - \frac{\mu}{u}m = 0,$$

$$gx + fy + \left(c - \frac{1}{u}\right)z - \frac{\mu}{u}n = 0.$$

Also $$lx + my + nz = 0.$$

Eliminate the ratios $x:y:z:-\mu/u$ between these four equations. Then

$$\begin{vmatrix} a - \dfrac{1}{u} & h & g & l \\[2mm] h & b - \dfrac{1}{u} & f & m \\[2mm] g & f & c - \dfrac{1}{u} & n \\[2mm] l & m & n & 0 \end{vmatrix} = 0.$$

This is, on expansion, a quadratic equation in $\frac{1}{u}$, and therefore in u, whose roots determine the maximum and minimum values of u. The lengths of the axes may thus be found.

The next illustration will help to consolidate what has been done, and also to expand certain details.

ILLUSTRATION 3. *To find the maxima and minima, for* NON-ZERO *values of* x, y, z, *of the function*

$$u \equiv xyz,$$

under the condition $\qquad x + y + z = 1.$

At a turning value, $du = 0$, so that

$$yz\,dx + zx\,dy + xy\,dz = 0.$$

By the equation of condition,

$$dx + dy + dz = 0.$$

Multiply these two equations among the differentials by 1, λ respectively, and add. Then choose λ so that the coefficient of dx vanishes; that is, so that

$$yz + \lambda = 0.$$

Then $\qquad [zx + \lambda]\,dy + [xy + \lambda]\,dz = 0.$

Now the *three* variables x, y, z are subject to *one* condition; their three differentials are therefore subject to one linear condition, so that *two* differentials may be given any arbitrarily assigned values. In particular, dy, dz may be assigned arbitrarily. Hence the coefficients of dy, dz in the equation last written must both be zero, so that

$$zx + \lambda = 0,$$

$$xy + \lambda = 0.$$

To summarize, the *four* equations

$$x + y + z = 1,$$

$$yz + \lambda = 0,$$

$$zx + \lambda = 0,$$

$$xy + \lambda = 0,$$

determine the four unknowns x, y, z, λ.

Multiply the last three equations by x, y, z, and add. Then

$$3xyz + \lambda(x + y + z) = 0,$$

or $\qquad 3u + \lambda = 0,$

or $\qquad \lambda = -3u.$

Thus $\qquad yz = 3u,$

$$zx = 3u,$$

$$xy = 3u.$$

Multiply corresponding sides of these three equations; then

$$x^2y^2z^2 = 27u^3,$$

or
$$u^2 = 27u^3.$$

Now u is not zero, under the conditions of the problem (x, y, z non-zero). Hence
$$u = \tfrac{1}{27}.$$

We may extend this illustration to show how to settle whether such points are maxima or minima.

It is easy to prove that there is only one such point, namely,

$$x = y = z = \tfrac{1}{3}.$$

Consider, then, a near point

$$x = \tfrac{1}{3}+p, \quad y = \tfrac{1}{3}+q, \quad z = \tfrac{1}{3}+r,$$

where p, q, r are small. Since $x+y+z = 1$, we have

$$p+q+r = 0.$$

Now $u = xyz$

$$= (\tfrac{1}{3}+p)\,(\tfrac{1}{3}+q)\,(\tfrac{1}{3}+r)$$

$$= \tfrac{1}{27}+\tfrac{1}{9}(p+q+r)+\tfrac{1}{3}(qr+rp+pq)+pqr.$$

Suppose that p, q, r are so small that the product pqr may be neglected. Then

$$u = \tfrac{1}{27}+\tfrac{1}{3}\{\tfrac{1}{2}(p+q+r)^2-\tfrac{1}{2}(p^2+q^2+r^2)\}$$

$$= \tfrac{1}{27}-\tfrac{1}{6}(p^2+q^2+r^2).$$

Hence u has its locally *greatest* value at $(\tfrac{1}{3}, \tfrac{1}{3}, \tfrac{1}{3})$, and the value $\tfrac{1}{27}$ is therefore a *maximum*.

We now apply these ideas more generally.

To locate the maxima or minima of the function

$$u \equiv f(x_1, x_2, ..., x_n)$$

of the n variables $x_1, x_2, ..., x_n$ subject to the m relations ($m < n$)

$$\phi_1(x_1, x_2, ..., x_n) = 0,$$

$$\phi_2(x_1, x_2, ..., x_n) = 0,$$

$$\cdots\cdots\cdots\cdots\cdots\cdots$$

$$\phi_m(x_1, x_2, ..., x_n) = 0.$$

At a turning value, $du = 0$, so that

$$\frac{\partial f}{\partial x_1} dx_1 + \frac{\partial f}{\partial x_2} dx_2 + \ldots + \frac{\partial f}{\partial x_n} dx_n = 0.$$

By the equations of condition,

$$\frac{\partial \phi_1}{\partial x_1} dx_1 + \frac{\partial \phi_1}{\partial x_2} dx_2 + \ldots + \frac{\partial \phi_1}{\partial x_n} dx_n = 0,$$

$$\cdots\cdots\cdots\cdots\cdots\cdots\cdots\cdots\cdots$$

$$\frac{\partial \phi_m}{\partial x_1} dx_1 + \frac{\partial \phi_m}{\partial x_2} dx_2 + \ldots + \frac{\partial \phi_m}{\partial x_n} dx_n = 0.$$

Multiply these $m+1$ equations among the differentials by $1, \lambda_1, \lambda_2, \ldots, \lambda_m$ respectively, and add. Then choose the multipliers λ in such a way that the coefficients of $dx_1, dx_2, \ldots, dx_m$ all vanish; that is, so that

$$\frac{\partial f}{\partial x_1} + \lambda_1 \frac{\partial \phi_1}{\partial x_1} + \ldots + \lambda_m \frac{\partial \phi_m}{\partial x_1} = 0,$$

$$\cdots\cdots\cdots\cdots\cdots\cdots\cdots\cdots\cdots$$

$$\frac{\partial f}{\partial x_m} + \lambda_1 \frac{\partial \phi_1}{\partial x_m} + \ldots + \lambda_m \frac{\partial \phi_m}{\partial x_m} = 0.$$

Then $\left[\dfrac{\partial f}{\partial x_{m+1}} + \lambda_1 \dfrac{\partial \phi_1}{\partial x_{m+1}} + \ldots + \lambda_m \dfrac{\partial \phi_m}{\partial x_{m+1}} \right] dx_{m+1} + \ldots$

$$+ \left[\frac{\partial f}{\partial x_n} + \lambda_1 \frac{\partial \phi_1}{\partial x_n} + \ldots + \lambda_m \frac{\partial \phi_m}{\partial x_n} \right] dx_n = 0.$$

Now the n variables $x_1, \ldots, x_n$ are subject to m conditions; their n differentials are therefore subject to m linear conditions, so that $n - m$ differentials may be given arbitrarily assigned values. In particular, $dx_{m+1}, dx_{m+2}, \ldots, dx_n$ may be assigned arbitrarily. Hence the coefficients of $dx_{m+1}, \ldots, dx_n$ in the equation last written must be zero, so that

$$\frac{\partial f}{\partial x_{m+1}} + \lambda_1 \frac{\partial \phi_1}{\partial x_{m+1}} + \ldots + \lambda_m \frac{\partial \phi_m}{\partial x_{m+1}} = 0,$$

$$\cdots\cdots\cdots\cdots\cdots\cdots\cdots\cdots\cdots$$

$$\frac{\partial f}{\partial x_n} + \lambda_1 \frac{\partial \phi_1}{\partial x_n} + \ldots + \lambda_m \frac{\partial \phi_m}{\partial x_n} = 0.$$

In all, then, we have the $m+n$ equations

$$\phi_i(x_1, x_2, \ldots, x_n) = 0 \quad (i = 1, \ldots, m),$$

$$\frac{\partial f}{\partial x_j} + \lambda_1 \frac{\partial \phi_1}{\partial x_j} + \ldots + \lambda_m \frac{\partial \phi_m}{\partial x_j} = 0 \quad (j = 1, \ldots, n)$$

to determine the $m+n$ unknowns $x_1, \ldots, x_n$, $\lambda_1, \ldots, \lambda_m$. This can, *in general*, be effected, and the value of u ascertained.

Note. The method breaks down if all the determinants formed by taking m columns of the 'matrix'

$$\frac{\partial \phi_1}{\partial x_1}, \quad \frac{\partial \phi_1}{\partial x_2}, \quad \ldots, \quad \frac{\partial \phi_1}{\partial x_n},$$

$$\frac{\partial \phi_2}{\partial x_1}, \quad \frac{\partial \phi_2}{\partial x_2}, \quad \ldots, \quad \frac{\partial \phi_2}{\partial x_n},$$

$$\ldots \ldots \ldots \ldots \ldots \ldots \ldots \ldots \ldots$$

$$\frac{\partial \phi_m}{\partial x_1}, \quad \frac{\partial \phi_m}{\partial x_2}, \quad \ldots, \quad \frac{\partial \phi_m}{\partial x_n},$$

should vanish. The equations of condition are not then independent. But the reader is unlikely to be troubled by this case, at any rate for the present.

REVISION EXAMPLES XI

University Level

1. Find the maximum and minimum values of

$$\frac{x+y-1}{x^2 + 2y^2 + 2}.$$

2. If $z = \left(\dfrac{p}{x}\right)^k + \left(\dfrac{x}{y}\right)^k + \left(\dfrac{y}{q}\right)^k$, where p, q, k are constants, prove that z has a stationary value when p, x, y, q are in geometrical progression.

3. The variable z is determined as a function of x, y by the equation
$$x^3 + y^3 + z^3 - 6xyz + 3 = 0.$$

Find the values of $\dfrac{\partial^2 z}{\partial x^2}$, $\dfrac{\partial^2 z}{\partial y^2}$, $\dfrac{\partial^2 z}{\partial x \, \partial y}$ when $x = y = z = 1$, and prove that these values of x, y, z make the function

$$z^2 + 2xy$$

a minimum.

4. Find the minimum value of

$$x^2 + y^2 + (ax + by + c)^2.$$

5. Find all the stationary values of the function

$$\frac{x}{y}+\frac{y}{x}-\frac{(x-y)^2}{a^2}$$

of the variables x, y; discuss how their nature depends on the value of the parameter a. Consider in particular the points at which $x = y = \pm a$.

6. Find the greatest volume of a rectangular parallelepiped which can be placed inside an ellipsoid of semi-axes a, b, c with its edges parallel to the axes.

[The edges $2x$, $2y$, $2z$ of the box are subject to the condition

$$\frac{x^2}{a^2}+\frac{y^2}{b^2}+\frac{z^2}{c^2} = 1.]$$

7. Find the maximum and minimum values, for real values of x, y, z, of the quantity $x^2+y^2+z^2$ subject to the conditions that

$$lx+my+nz = 0,$$
$$ax^2+by^2+cz^2 = 1,$$

where a, b, c are positive and l, m, n are real.

Verify that the values determined are real and positive.

8. Find the maximum and minimum values of the function

$$u \equiv x^3+y^3+z^3,$$

where x, y, z are connected by the relations

$$x+y+z = a,$$
$$x^2+y^2+z^2 = a^2.$$

9. If $\qquad ax+by+cz = 1 \qquad (a, b, c \text{ positive})$

show that the values of x, y, z for which

$$\frac{1}{x}+\frac{1}{y}+\frac{1}{z}$$

is stationary are given by

$$ax^2 = by^2 = cz^2.$$

Show that this is a true maximum or minimum if $xyz > 0$.

10. Use Lagrange's method of undetermined multipliers to show that the triangle of maximum area which can be inscribed in a given circle is equilateral.

11. The sum of the twelve edges of a rectangular block is k; the sum of the areas of the six faces is $\frac{1}{28}k^2$. Prove that, when the excess of the volume of the block over that of a cube, whose edge is equal to the least edge of the block, is greatest, the least edge is $\frac{1}{28}k$; and find the other edges.

12. Find the greatest and least distances from the origin of a point on the surface

$$\left(\frac{x}{a}\right)^p + \left(\frac{y}{b}\right)^p + \left(\frac{z}{c}\right)^p = 1,$$

where a, b, c are fixed positive numbers and p is a fixed even integer greater than 2.

[The square of the distance is $x^2 + y^2 + z^2$.]

13. Prove that, if α, β, γ are positive, there exists a shape of triangle for which the maximum value of

$$\sin^\alpha A \,\sin^\beta B \,\sin^\gamma C$$

is attained; and that for this triangle

$$\tan^2 A = \frac{\alpha(\alpha+\beta+\gamma)}{\beta\gamma}, \quad \tan^2 B = \frac{\beta(\alpha+\beta+\gamma)}{\gamma\alpha}, \quad \tan^2 C = \frac{\gamma(\alpha+\beta+\gamma)}{\alpha\beta}.$$

14. Find the stationary values of

$$y^2 + 4z^2 - 4yz - 2zx - 2xy$$

subject to $2x^2 + 3y^2 + 6z^2 = 1.$

15. Show that, if x, y, z are connected by the relations

$$x^2 + y^2 + z^2 = 1,$$
$$lx + my + nz = 0,$$

and a, b, c are not all equal, then the extreme values of the function

$$V \equiv ax^2 + by^2 + cz^2$$

are the roots of the equation

$$\frac{l^2}{v-a} + \frac{m^2}{v-b} + \frac{n^2}{v-c} = 0.$$

16. On the surface given by

$$x^2 + y^2 + z^2 - 2x + 2y + 6z + 9 = 0,$$

find the stationary points of the function $x^2 + y^2 + z^2 - yz - zx - xy$, and investigate their nature.

17. The variables x, y, z are connected by the relations

$$\phi(x, y, z) = 0, \quad \psi(x, y, z) = 0.$$

Prove that a necessary condition for $f(x, y, z)$ to have a stationary value at (x_0, y_0, z_0) is that

$$\begin{vmatrix} \dfrac{\partial f}{\partial x} & \dfrac{\partial f}{\partial y} & \dfrac{\partial f}{\partial z} \\[2mm] \dfrac{\partial \phi}{\partial x} & \dfrac{\partial \phi}{\partial y} & \dfrac{\partial \phi}{\partial z} \\[2mm] \dfrac{\partial \psi}{\partial x} & \dfrac{\partial \psi}{\partial y} & \dfrac{\partial \psi}{\partial z} \end{vmatrix}$$

should vanish at (x_0, y_0, z_0), where f, ϕ, ψ are assumed to be differentiable.

Prove that, if $\quad x + y + z = -xyz = 1$,

then there are three points (x, y, z) at which $x^2 + y^2 + z^2$ has a stationary value, and show that they are minima.

18. If $\qquad f(x, y) \equiv x^3 - 3xy^2 + 18y$,

where $\qquad 3x^2y - y^3 - 6x = 0$,

prove that the values of x, y which make $f(x, y)$ a maximum or minimum are $x = y = \pm \sqrt{3}$.

19. Investigate the character at the points $(-5, 4, 4)$, $(1, 1, 1)$ of the function $x^2 + y^2 + z^2$, where

$$x + y + z = 3, \quad x^3 + y^3 + z^3 = 3.$$

20. Determine the minimum value of

$$ax^2 + by^2 + cz^2,$$

where a, b, c are positive constants and the variables are restricted by the relationship $x + y + z = 1$.

Hence, or otherwise, determine the minimum value of

$$px^2 + qy^2 + rz^2 + 2yz + 2zx + 2xy$$

subject to the same restriction, where p, q, r are constants, greater than unity.

21. Show that the minimum value of

$$(a^2x^2 + b^2y^2 + c^2z^2)/x^2y^2z^2,$$

where $ax^2 + by^2 + cz^2 = 1$, with a, b, c positive, is given by

$$x^2 = \frac{u}{2a(u+a)}, \quad y^2 = \frac{u}{2b(u+b)}, \quad z^2 = \frac{u}{2c(u+c)},$$

where u is the positive root of the equation

$$u^3 - (bc + ca + ab)\,u - 2abc = 0.$$

22. If r denotes the distance from the origin to a point on the curve in which the plane $lx + my + nz = 0$ meets the surface

$$(x^2 + y^2 + z^2)^2 = a^2x^2 + b^2y^2 + c^2z^2,$$

prove that the non-zero maximum and minimum values of r^2 are roots of the equation

$$\frac{l^2}{a^2 - r^2} + \frac{m^2}{b^2 - r^2} + \frac{n^2}{c^2 - r^2} = 0.$$

23. If m, n, p are given positive numbers, find positive numbers x, y, z, whose sum is a constant A, such that the function $x^m y^n z^p$ is a maximum.

CHAPTER XVI

JACOBIANS

1. Introductory example. Suppose that two variables x, y are expressed as functions of u, v in the form

$$x = u^2 + 2v,$$

$$y = u + v.$$

Then u, v can also be expressed in terms of x, y; for

$$x - 2y = u^2 - 2u,$$

so that $\qquad (u-1)^2 = x - 2y + 1,$

or $\qquad u = 1 \pm \sqrt{(x - 2y + 1)},$

and so $\qquad v = y - 1 \mp \sqrt{(x - 2y + 1)}.$

The given relations express x, y *uniquely* as functions of u, v; but there are *two* sets of relations to express u, v as functions of x, y, namely, EITHER

$$\begin{cases} u = 1 + \sqrt{(x - 2y + 1)}, \\ v = y - 1 - \sqrt{(x - 2y + 1)}, \end{cases}$$

OR, alternatively,

$$\begin{cases} u = 1 - \sqrt{(x - 2y + 1)}, \\ v = y - 1 + \sqrt{(x - 2y + 1)}. \end{cases}$$

If, as is natural, we now make the assumption that u, v are to be *continuous*, then in any particular problem we must keep to one or other of these two alternatives and not move between them. For example, if we choose the FIRST pair of expressions when $x = 6$, $y = 3$, we have

$$\begin{cases} u = 1 + \sqrt{(6 - 6 + 1)} \quad = 2, \\ v = 3 - 1 - \sqrt{(6 - 6 + 1)} = 1; \end{cases}$$

and if we then proceeded to take the SECOND for the near values $x = 6 \cdot 21, y = 3$, we should have

$$\begin{cases} u = 1 - \sqrt{(6 \cdot 21 - 6 + 1)} \quad = 1 - 1 \cdot 1 = -0 \cdot 1, \\ v = 3 - 1 + \sqrt{(6 \cdot 21 - 6 + 1)} = 2 + 1 \cdot 1 = \quad 3 \cdot 1. \end{cases}$$

The pairs $\qquad \begin{cases} u = 2 \\ v = 1 \end{cases}$ and $\begin{cases} u = -0 \cdot 1 \\ v = \quad 3 \cdot 1 \end{cases}$

clearly reveal a discontinuity from which the pairs

$$\begin{cases} u = 2 \\ v = 1 \end{cases} \text{ and } \begin{cases} u = 2\cdot1 \\ v = 0\cdot9 \end{cases}$$

(formed by keeping to the FIRST expressions) would not suffer.

We can therefore express u, v *uniquely* as continuous functions of x, y by using EITHER the relations

$$\begin{cases} u = 1 + \sqrt{(x - 2y + 1)}, \\ v = y - 1 - \sqrt{(x - 2y + 1)}, \end{cases}$$

OR the relations

$$\begin{cases} u = 1 - \sqrt{(x - 2y + 1)}, \\ v = y - 1 + \sqrt{(x - 2y + 1)}. \end{cases}$$

There is, however, one condition under which this uniqueness of expression breaks down. If the values x, y are chosen so that

$$x - 2y + 1 = 0,$$

we obtain without ambiguity the solution

$$u = 1, \quad v = y - 1$$

and we can, for values of x, y near to those so chosen, take either of the alternative forms of expression while retaining continuity. For example, if $x = 7$, $y = 4$, then either of the pairs

$$\begin{cases} u = 1 + \sqrt{(7 - 8 + 1)} \\ v = 4 - 1 - \sqrt{(7 - 8 + 1)} \end{cases} \text{ or } \begin{cases} u = 1 - \sqrt{(7 - 8 + 1)} \\ v = 4 - 1 + \sqrt{(7 - 8 + 1)} \end{cases}$$

gives

$$u = 1, \quad v = 3.$$

Moreover, the near values $x = 6\cdot9801$, $y = 3\cdot99$ give, for the first alternative,

$$\begin{cases} u = 1 + \sqrt{(6\cdot9801 - 7\cdot98 + 1)} \quad\quad = 1 + \sqrt{(0\cdot0001)} = 1\cdot01, \\ v = 3\cdot99 - 1 - \sqrt{(6\cdot9801 - 7\cdot98 + 1)} = 2\cdot99 - 0\cdot01 \quad = 2\cdot98, \end{cases}$$

and, for the second,

$$\begin{cases} u = 1 - 0\cdot01 \quad = 0\cdot99, \\ v = 2\cdot99 + 0\cdot01 = 3. \end{cases}$$

Each of the alternatives $(1\cdot01, 2\cdot98)$ and $(0\cdot99, 3)$ maintains continuity with the pair $(1, 3)$.

We may give a geometrical formulation of what we have just indicated, taking x, y as the coordinates of a point in a plane (fig. 113) in which the line $x - 2y + 1 = 0$ is drawn.

Consider, say, the solution

$$\begin{cases} u = 1 + \sqrt{(x - 2y + 1)}, \\ v = y - 1 - \sqrt{(x - 2y + 1)}. \end{cases}$$

When we have decided on these formulae, the expressions for u, v in terms of x, y are unique and continuous. It is implicit, though, that $x - 2y + 1$ is to be positive, so that the point (x, y) always lies

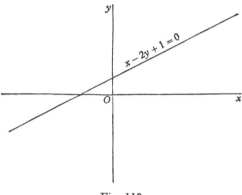

Fig. 113.

'below' the line $x - 2y + 1 = 0$. If, then, (x, y) moves continuously along any curve in the 'lower' part of the plane, u and v vary continuously. But it is possible to *move over* to the alternative formulae

$$\begin{cases} u = 1 - \sqrt{(x - 2y + 1)}, \\ v = y - 1 + \sqrt{(x - 2y + 1)} \end{cases}$$

at any point where the path of the point (x, y) meets the line $x - 2y + 1 = 0$. For *either* choice, the values of u, v are continuous over the 'lower' part of the plane, even at the line.

To summarize: suppose that, at any one point (x_1, y_1) in the 'lower' part of the (x, y)-plane, the choice of solution

$$\begin{cases} u = 1 + \sqrt{(x_1 - 2y_1 + 1)}, \\ v = y_1 - 1 - \sqrt{(x_1 - 2y_1 + 1)} \end{cases}$$

has been made. The values of u, v at any other point (x_2, y_2) can be obtained by allowing u, v to vary continuously with changes in x, y such that the point (x, y) moves by a continuous curve (in the 'lower' part) from (x_1, y_1) to (x_2, y_2). If the points (x_1, y_1), (x_2, y_2) are joined by a curve not meeting the line $x - 2y + 1 = 0$, the values at (x_2, y_2) are

$$u = 1 + \sqrt{(x_2 - 2y_2 + 1)}, \quad v = y_2 - 1 - \sqrt{(x_2 - 2y_2 + 1)}.$$

If, however, the points are joined by a curve meeting the line, a change can be made if desired as the point (x, y), moving along the continuous curve, meets the line. Once the decision has been made, the formulae to be used for u, v remain unique and must be kept.

The only ambiguity in the choice of functions lies at the line $x - 2y + 1 = 0$, and it is worthy of explicit remark that u, v are genuinely two-valued functions of x, y there. If the point (x_1, y_1) lies ON that line, then *each* of the pairs of values

$$\begin{cases} u = 1 + \sqrt{(x_2 - 2y_2 + 1)} \\ v = y_2 - 1 - \sqrt{(x_2 - 2y_2 + 1)} \end{cases} \text{ and } \begin{cases} u = 1 - \sqrt{(x_2 - 2y_2 + 1)} \\ v = y_2 - 1 + \sqrt{(x_2 - 2y_2 + 1)} \end{cases}$$

may be obtained at (x_2, y_2) while retaining continuity as the point (x, y) moves along a continuous curve from (x_1, y_1) to (x_2, y_2).

We now turn our attention to the differentials of the given functions

$$x = u^2 + 2v, \quad y = u + v,$$

namely,

$$dx = 2u\, du + 2\, dv, \quad dy = du + dv.$$

Solving these for du, dv by the usual process of elimination, we have the relations

$$2(u - 1)\, du = dx - 2\, dy,$$

$$2(u - 1)\, dv = -dx + 2u\, dy,$$

leading to the partial differential coefficients

$$\frac{\partial u}{\partial x} = \frac{1}{2(u - 1)}, \quad \frac{\partial u}{\partial y} = -\frac{1}{u - 1},$$

$$\frac{\partial v}{\partial x} = -\frac{1}{2(u - 1)}, \quad \frac{\partial v}{\partial y} = \frac{u}{u - 1}.$$

There is, however, one case where this analysis breaks down. When

$$u = 1,$$

the relations for du, dv give simply the equations

$$0 = dx - 2\, dy,$$

$$0 = -dx + 2\, dy$$

which do not involve du, dv at all, and the coefficients $\dfrac{\partial u}{\partial x}, \dfrac{\partial u}{\partial y}, \dfrac{\partial v}{\partial x}, \dfrac{\partial v}{\partial y}$ do not exist. Moreover, this relation $u = 1$ is exactly equivalent to the earlier relation

$$x - 2y + 1 = 0.$$

In other words, *the differential coefficients* $\dfrac{\partial u}{\partial x}, \dfrac{\partial u}{\partial y}, \dfrac{\partial v}{\partial x}, \dfrac{\partial v}{\partial y}$ *do not exist at precisely those points* (*namely, those for which* $x - 2y + 1 = 0$) *where the expression of u, v in terms of x, y is not single-valued.*

It may be worth while to comment that the variables x, y are related at the 'doubtful' points for which $u = 1$, by the expression

$$x - 2y + 1 = 0;$$

and that the differentials dx, dy are then connected by the relation

$$dx - 2dy = 0.$$

Finally, let us generalize the functions to exhibit a clearer view of what is involved. Write

$$x = f(u, v), \quad y = g(u, v),$$

where $f(u, v)$, $g(u, v)$ are single-valued functions when the point (u, v) lies within a certain region of the (u, v)-plane. The differential coefficients $\dfrac{\partial u}{\partial x}, \dfrac{\partial u}{\partial y}, \dfrac{\partial v}{\partial x}, \dfrac{\partial v}{\partial y}$ may, in general, be calculated by an appeal to the differentials of the given relations in the form

$$dx = \frac{\partial f}{\partial u}\,du + \frac{\partial f}{\partial v}\,dv,$$

$$dy = \frac{\partial g}{\partial u}\,du + \frac{\partial g}{\partial v}\,dv;$$

for elimination of dv, du in turn gives the equations

$$\left(\frac{\partial f}{\partial u}\frac{\partial g}{\partial v} - \frac{\partial f}{\partial v}\frac{\partial g}{\partial u}\right) du = \frac{\partial g}{\partial v}\,dx - \frac{\partial f}{\partial v}\,dy,$$

$$\left(\frac{\partial f}{\partial u}\frac{\partial g}{\partial v} - \frac{\partial f}{\partial v}\frac{\partial g}{\partial u}\right) dv = -\frac{\partial g}{\partial u}\,dx + \frac{\partial f}{\partial u}\,dy,$$

and the partial differential coefficients follow, EXCEPT for values of u, v such that

$$\frac{\partial f}{\partial u}\frac{\partial g}{\partial v} - \frac{\partial f}{\partial v}\frac{\partial g}{\partial u} = 0.$$

The left-hand side of this equation is the determinant

$$\begin{vmatrix} \dfrac{\partial f}{\partial u} & \dfrac{\partial f}{\partial v} \\[2ex] \dfrac{\partial g}{\partial u} & \dfrac{\partial g}{\partial v} \end{vmatrix}$$

of the coefficients of du, dv in the equations for dx, dy, and is of great importance. It is the function whose vanishing in the particular case $x = u^2 + 2v$, $y = u + v$ gave the relation $u = 1$ to locate the points where $\dfrac{\partial u}{\partial x}$, $\dfrac{\partial u}{\partial y}$, $\dfrac{\partial v}{\partial x}$, $\dfrac{\partial v}{\partial y}$ did not exist. With generalizations to larger numbers of variables, this determinant must now be studied more closely.

EXAMPLE I

Consider similarly the relations

$$x = r \cos \theta, \quad y = r \sin \theta,$$

and show that

(i) the equations can be solved to give r, θ uniquely (to within multiples of 2π for θ) in terms of x, y *except* when $r = 0$, the value of θ being then arbitrary;

(ii) $$\frac{\partial x}{\partial r}\frac{\partial y}{\partial \theta} - \frac{\partial x}{\partial \theta}\frac{\partial y}{\partial r} = 0$$

when $r = 0$;

(iii) the partial differential coefficients $\dfrac{\partial r}{\partial x}$, $\dfrac{\partial r}{\partial y}$, $\dfrac{\partial \theta}{\partial x}$, $\dfrac{\partial \theta}{\partial y}$ do not exist when $r = 0$.

2. The Jacobian defined. Taking, first, the case of three variables as an illustration, suppose that

$$f(u, v, w), \quad g(u, v, w), \quad h(u, v, w)$$

are three functions of the variables u, v, w. The determinant

$$J \equiv \begin{vmatrix} \dfrac{\partial f}{\partial u} & \dfrac{\partial f}{\partial v} & \dfrac{\partial f}{\partial w} \\[2ex] \dfrac{\partial g}{\partial u} & \dfrac{\partial g}{\partial v} & \dfrac{\partial g}{\partial w} \\[2ex] \dfrac{\partial h}{\partial u} & \dfrac{\partial h}{\partial v} & \dfrac{\partial h}{\partial w} \end{vmatrix}$$

is called the *Jacobian* of the three functions, and is often denoted by the notation

$$J \equiv \frac{\partial(f, g, h)}{\partial(u, v, w)}.$$

A similar definition provides us with the Jacobian

$$\frac{\partial(x_1, x_2, \ldots, x_n)}{\partial(u_1, u_2, \ldots, u_n)}$$

of the n functions $x_1, x_2, \ldots, x_n$ of the n variables $u_1, u_2, \ldots, u_n$, in the form

$$J \equiv \begin{vmatrix} \dfrac{\partial x_1}{\partial u_1} & \dfrac{\partial x_1}{\partial u_2} & \cdots & \dfrac{\partial x_1}{\partial u_n} \\ \dfrac{\partial x_2}{\partial u_1} & \dfrac{\partial x_2}{\partial u_2} & \cdots & \dfrac{\partial x_2}{\partial u_n} \\ \cdots\cdots\cdots\cdots\cdots\cdots\cdots\cdots \\ \dfrac{\partial x_n}{\partial u_1} & \dfrac{\partial x_n}{\partial u_2} & \cdots & \dfrac{\partial x_n}{\partial u_n} \end{vmatrix}.$$

We shall, in fact, usually deal with the general number n, as the restriction to the number 3 affords no real simplification.

ILLUSTRATION 1. *The variables x, y, z are expressed as functions of the variables r, θ, ϕ by means of the relations*

$$x = r \sin \theta \cos \phi, \quad y = r \sin \theta \sin \phi, \quad z = r \cos \theta.$$

To evaluate the Jacobian $\dfrac{\partial(x, y, z)}{\partial(r, \theta, \phi)}$.

By definition

$$\frac{\partial(x, y, z)}{\partial(r, \theta, \phi)} = \begin{vmatrix} \dfrac{\partial x}{\partial r} & \dfrac{\partial x}{\partial \theta} & \dfrac{\partial x}{\partial \phi} \\ \dfrac{\partial y}{\partial r} & \dfrac{\partial y}{\partial \theta} & \dfrac{\partial y}{\partial \phi} \\ \dfrac{\partial z}{\partial r} & \dfrac{\partial z}{\partial \theta} & \dfrac{\partial z}{\partial \phi} \end{vmatrix}$$

$$= \begin{vmatrix} \sin \theta \cos \phi & r \cos \theta \cos \phi & -r \sin \theta \sin \phi \\ \sin \theta \sin \phi & r \cos \theta \sin \phi & r \sin \theta \cos \phi \\ \cos \theta & -r \sin \theta & 0 \end{vmatrix}.$$

Hence, by elementary expansion of the determinant,

$$\frac{\partial(x, y, z)}{\partial(r, \theta, \phi)} = r^2 \sin \theta.$$

EXAMPLES II

1. If $x = r\cos\theta$, $y = r\sin\theta$, evaluate $\dfrac{\partial(x,y)}{\partial(r,\theta)}$.

2. If $x = e^u\cos\phi$, $y = e^u\sin\phi$, evaluate $\dfrac{\partial(x,y)}{\partial(u,\phi)}$.

3. If $x = r\cosh\theta$, $y = r\sinh\theta$, evaluate $\dfrac{\partial(x,y)}{\partial(r,\theta)}$.

4. If $u = x^2+y^2+z^2$, $v = (x+y+z)^2$, $w = yz+zx+xy$, evaluate $\dfrac{\partial(u,v,w)}{\partial(x,y,z)}$.

5. If $u = e^t\sin\theta\cos\phi$, $v = e^t\sin\theta\sin\phi$, $w = e^t\cos\theta$, evaluate $\dfrac{\partial(u,v,w)}{\partial(t,\theta,\phi)}$.

3. The chain rule for Jacobians. *To prove that, if* $x_1, x_2, ..., x_n$ *are functions of* $u_1, u_2, ..., u_n$, *which are themselves functions of* $\xi_1, \xi_2, ..., \xi_n$, *then*

$$\frac{\partial(x_1, x_2, ..., x_n)}{\partial(u_1, u_2, ..., u_n)}\frac{\partial(u_1, u_2, ..., u_n)}{\partial(\xi_1, \xi_2, ..., \xi_n)} = \frac{\partial(x_1, x_2, ..., x_n)}{\partial(\xi_1, \xi_2, ..., \xi_n)}.$$

The proof rests on the formula for the product of two determinants, namely, that

$$\begin{vmatrix} a_{11} & a_{12} & \cdots & a_{1n} \\ a_{21} & a_{22} & \cdots & a_{2n} \\ \cdots\cdots\cdots\cdots\cdots \\ a_{n1} & a_{n2} & \cdots & a_{nn} \end{vmatrix} \times \begin{vmatrix} b_{11} & b_{12} & \cdots & b_{1n} \\ b_{21} & b_{22} & \cdots & b_{2n} \\ \cdots\cdots\cdots\cdots\cdots \\ b_{n1} & b_{n2} & \cdots & b_{nn} \end{vmatrix} = \begin{vmatrix} c_{11} & c_{12} & \cdots & c_{1n} \\ c_{21} & c_{22} & \cdots & c_{2n} \\ \cdots\cdots\cdots\cdots\cdots \\ c_{n1} & c_{n2} & \cdots & c_{nn} \end{vmatrix},$$

where $$c_{ij} = \sum_{\lambda=1}^{n} a_{i\lambda} b_{\lambda j}.$$

[The reader may confine his attention to the case $n = 3$ if that is more familiar.]

If we write $$a_{ij} \equiv \frac{\partial x_i}{\partial u_j}, \quad b_{ij} \equiv \frac{\partial u_i}{\partial \xi_j},$$

so that the two determinants to be multiplied are simply

$$\frac{\partial(x_1, x_2, ..., x_n)}{\partial(u_1, u_2, ..., u_n)}, \quad \frac{\partial(u_1, u_2, ..., u_n)}{\partial(\xi_1, \xi_2, ..., \xi_n)},$$

then
$$c_{ij} \equiv \sum_{\lambda=1}^{n} \frac{\partial x_i}{\partial u_\lambda} \frac{\partial u_\lambda}{\partial \xi_j}$$

$$= \frac{\partial x_i}{\partial \xi_j} \quad \text{(p. 27)}.$$

Hence the product determinant is $\dfrac{\partial(x_1, x_2, \ldots, x_n)}{\partial(\xi_1, \xi_2, \ldots, \xi_n)}$, as we were to prove.

ILLUSTRATION 2. *To verify the chain rule, in the form*

$$\frac{\partial(u, v)}{\partial(x, y)} \frac{\partial(x, y)}{\partial(r, \theta)} = \frac{\partial(u, v)}{\partial(r, \theta)},$$

for the relations $\quad u = x^2 + y^2, \quad v = 2xy,$

$$x = r \cos \theta, \quad y = r \sin \theta.$$

We have
$$\frac{\partial(u, v)}{\partial(x, y)} = \begin{vmatrix} 2x & 2y \\ 2y & 2x \end{vmatrix} = 4(x^2 - y^2)$$

$$= 4r^2 \cos 2\theta,$$

and
$$\frac{\partial(x, y)}{\partial(r, \theta)} = \begin{vmatrix} \cos \theta & \sin \theta \\ -r \sin \theta & r \cos \theta \end{vmatrix}$$

$$= r.$$

Thus
$$\frac{\partial(u, v)}{\partial(x, y)} \frac{\partial(x, y)}{\partial(r, \theta)} = 4r^3 \cos 2\theta.$$

Now
$$u = r^2 \cos^2 \theta + r^2 \sin^2 \theta$$

$$= r^2,$$

$$v = 2r \cos \theta . r \sin \theta$$

$$= r^2 \sin 2\theta.$$

Thus
$$\frac{\partial(u, v)}{\partial(r, \theta)} = \begin{vmatrix} 2r & 0 \\ 2r \sin 2\theta & 2r^2 \cos 2\theta \end{vmatrix}$$

$$= 4r^3 \cos 2\theta.$$

The formula is therefore verified.

4. The 'reciprocal' theorem. *To prove that*

$$\frac{\partial(u_1, u_2, \ldots, u_n)}{\partial(x_1, x_2, \ldots, x_n)} = 1 \bigg/ \frac{\partial(x_1, x_2, \ldots, x_n)}{\partial(u_1, u_2, \ldots, u_n)}.$$

In the preceding paragraph, identify $\xi_1, \xi_2, ..., \xi_n$ with $x_1, x_2, ..., x_n$. Then

$$\frac{\partial(x_1, x_2, ..., x_n)}{\partial(u_1, u_2, ..., u_n)} \times \frac{\partial(u_1, u_2, ..., u_n)}{\partial(x_1, x_2, ..., x_n)}$$

$$= \frac{\partial(x_1, x_2, ..., x_n)}{\partial(x_1, x_2, ..., x_n)}$$

$$= \begin{vmatrix} 1 & 0 & 0 & 0 & \cdots \\ 0 & 1 & 0 & 0 & \cdots \\ 0 & 0 & 1 & 0 & \cdots \\ 0 & 0 & 0 & 1 & \cdots \\ \cdots\cdots\cdots\cdots\cdots \end{vmatrix}$$

$$= 1,$$

and the result follows.

ILLUSTRATION 3. *The variables r, θ, ϕ are functions of x, y, z according to the relations*

$$x = r \sin \theta \cos \phi, \quad y = r \sin \theta \sin \phi, \quad z = r \cos \theta.$$

To evaluate the Jacobian $\dfrac{\partial(r, \theta, \phi)}{\partial(x, y, z)}$.

The expression of r, θ, ϕ directly in terms of x, y, z is awkward; the subsequent differentiations are worse. We therefore use the result (p. 96)

$$\frac{\partial(x, y, z)}{\partial(r, \theta, \phi)} = r^2 \sin \theta,$$

and find at once that $\quad \dfrac{\partial(r, \theta, \phi)}{\partial(x, y, z)} = \dfrac{1}{r^2 \sin \theta}$.

REMARK. The importance of Jacobians centres in many ways round two fundamental properties which we are now to investigate.

First, they give a test to determine whether given functions are independent or connected by some 'functional relation'—the phrase being used in the sense that, for example, the functions

$$x = \sin^2 u + \cos^2 v, \quad y = \cos^2 u + \sin^2 v$$

are connected by the 'functional relation'

$$x + y = 2.$$

Secondly, they give an indication of the 'magnification' between corresponding (small) figures in, say, the (u, v)-plane and the (x, y)-plane or the (u, v, w)-space and the (x, y, z)-space. The analogy of the relation

$$\delta u = \frac{du}{dx} \delta x$$

between lengths δu, δx suggests itself at once.

The second of these properties will assume importance later (p. 132) when we come to study multiple integrals.

5. Dependent functions. *To prove that, if there is a functional relation connecting the n functions $x_1, x_2, ..., x_n$ of the n variables $u_1, u_2, ..., u_n$, then the Jacobian*

$$J \equiv \frac{\partial(x_1, x_2, ..., x_n)}{\partial(u_1, u_2, ..., u_n)}$$

has zero value everywhere.

Suppose that the functions $x_1, x_2, ..., x_n$ are connected by the relation
$$f(x_1, x_2, ..., x_n) = 0.$$
Differentiate with respect to $u_1, u_2, ..., u_n$ respectively. Then

$$\frac{\partial f}{\partial x_1}\frac{\partial x_1}{\partial u_1} + \frac{\partial f}{\partial x_2}\frac{\partial x_2}{\partial u_1} + ... + \frac{\partial f}{\partial x_n}\frac{\partial x_n}{\partial u_1} = 0,$$

$$\frac{\partial f}{\partial x_1}\frac{\partial x_1}{\partial u_2} + \frac{\partial f}{\partial x_2}\frac{\partial x_2}{\partial u_2} + ... + \frac{\partial f}{\partial x_n}\frac{\partial x_n}{\partial u_2} = 0,$$

$$\cdots\cdots\cdots\cdots\cdots\cdots\cdots\cdots\cdots\cdots\cdots$$

$$\frac{\partial f}{\partial x_1}\frac{\partial x_1}{\partial u_n} + \frac{\partial f}{\partial x_2}\frac{\partial x_2}{\partial u_n} + ... + \frac{\partial f}{\partial x_n}\frac{\partial x_n}{\partial u_n} = 0.$$

Now it is a familiar theorem in elimination that, if there are n relations,

$$a_1\xi_1 + a_2\xi_2 + ... + a_n\xi_n = 0,$$
$$b_1\xi_1 + b_2\xi_2 + ... + b_n\xi_n = 0,$$
$$\cdots\cdots\cdots\cdots\cdots\cdots\cdots\cdots\cdots$$
$$e_1\xi_1 + e_2\xi_2 + ... + e_n\xi_n = 0$$

among the n variables $\xi_1, \xi_2, ..., \xi_n$, then

$$\begin{vmatrix} a_1 & a_2 & ... & a_n \\ b_1 & b_2 & ... & b_n \\ \cdots & \cdots & \cdots & \cdots \\ e_1 & e_2 & ... & e_n \end{vmatrix} = 0,$$

provided that $\xi_1, \xi_2, \ldots, \xi_n$ are not themselves all zero. Applying this to the n equations obtained above, so as to eliminate

$$\frac{\partial f}{\partial x_1}, \quad \frac{\partial f}{\partial x_2}, \quad \ldots, \quad \frac{\partial f}{\partial x_n},$$

we have the required relation (interchanging rows and columns in the definition)

$$J \equiv \begin{vmatrix} \dfrac{\partial x_1}{\partial u_1} & \dfrac{\partial x_2}{\partial u_1} & \ldots & \dfrac{\partial x_n}{\partial u_1} \\[2mm] \dfrac{\partial x_1}{\partial u_2} & \dfrac{\partial x_2}{\partial u_2} & \ldots & \dfrac{\partial x_n}{\partial u_2} \\[1mm] \cdots\cdots\cdots\cdots\cdots\cdots\cdots \\[1mm] \dfrac{\partial x_1}{\partial u_n} & \dfrac{\partial x_2}{\partial u_n} & \ldots & \dfrac{\partial x_n}{\partial u_n} \end{vmatrix} = 0,$$

provided that $\dfrac{\partial f}{\partial x_1}, \dfrac{\partial f}{\partial x_2}, \ldots, \dfrac{\partial f}{\partial x_n}$ are not all zero; and the proviso is excluded by noting that, if it held, the given function $f(x_1, x_2, \ldots, x_n)$ would be independent of each of the functions $x_1, x_2, \ldots, x_n$—a case of little interest.

We state without proof the converse result:

If n functions of n variables have zero Jacobian everywhere, then there is a relation connecting them.

ILLUSTRATION 4. *To examine whether the three functions*

$$x + y + z, \quad x^2 + y^2 + z^2, \quad yz + zx + xy$$

are related.

The Jacobian of the functions is

$$J \equiv \begin{vmatrix} 1 & 1 & 1 \\ 2x & 2y & 2z \\ y+z & z+x & x+y \end{vmatrix},$$

and it is easy to prove that $J = 0$, so that the functions *are* related. (In fact,

$$(x + y + z)^2 - (x^2 + y^2 + z^2) - 2(yz + zx + xy) \equiv 0.)$$

EXAMPLES III

1. Verify that $\dfrac{\partial(x, y)}{\partial(u, v)} = 0$ for the functions

$$x \equiv \sin^2 u + \cos^2 v, \quad y = \cos^2 u + \sin^2 v.$$

2. Determine whether the functions

$$x^3 + y^3 + z^3, \quad xyz, \quad x + y + z$$

are related.

3. Determine whether the functions

$$x^3 + y^3 + z^3 - 3xyz, \quad x + y + z, \quad x^2 + y^2 + z^2$$

are related.

4. Determine whether there are values of the constant a for which the functions

$$x^3 + y^3 + z^3 + axyz, \quad x + y + z, \quad x^2 + y^2 + z^2 - yz - zx - xy$$

are related.

6. Ratio of areas under transformation. (Approximate theory.) The work now to be given is of fundamental importance, and exhibits one of the most characteristic features of Jacobian theory. Detailed discussion is difficult at the present stage, but the principles should be followed closely.

Suppose that the relations

$$u = f(x, y), \quad v = g(x, y)$$

define u, v as single-valued functions* of the variables x, y. Take (x, y) as coordinates in one plane and (u, v) as coordinates in another.

To each point P in the (x, y)-plane corresponds a point P' in the (u, v)-plane, and to any figure in the (x, y)-plane corresponds a figure in the (u, v)-plane. Our purpose is to prove that, *if A is the area of a small figure round the point (x_1, y_1), and if A' is the area of the corresponding small figure round (u_1, v_1), then, approximately,*

$$A' = \pm \frac{\partial(u_1, v_1)}{\partial(x_1, y_1)} A,$$

the sign being adjusted to make the right-hand side positive.

To do this we break the area A up into a number of small triangles; the figure A' is then broken correspondingly into a number of small figures which are also triangles (with sides *approximately* straight).

---

* It is understood that the relations can be solved to give x, y as *single-valued* functions of u, v, possibly for restricted ranges of values of the variables; for example, the relations $u = x^2$, $v = y^2$ can be solved uniquely in the form $x = \sqrt{u}$, $y = \sqrt{v}$ for points in the positive quadrants. See also the Note at the end of the next paragraph (p. 106).

Once the theorem is established for a small triangle, the general theorem will follow by addition (since all signs are adjusted to be positive), so we confine our attention to triangles.

Let us take, then, a small triangle with vertices $P_1(x_1, y_1)$, $P_2(x_2, y_2)$, $P_3(x_3, y_3)$, where

$$x_2 = x_1 + h, \quad y_2 = y_1 + k,$$
$$x_3 = x_1 + h', \quad y_3 = y_1 + k'.$$

It is known that the area of the triangle $P_1 P_2 P_3$ is A, where

$$2A = \pm \begin{vmatrix} x_1 & y_1 & 1 \\ x_2 & y_2 & 1 \\ x_3 & y_3 & 1 \end{vmatrix},$$

the sign being adjusted to make A positive. Thus

$$2A = \pm \begin{vmatrix} x_1 & y_1 & 1 \\ x_1+h & y_1+k & 1 \\ x_1+h' & y_1+k' & 1 \end{vmatrix}$$

$$= \pm \begin{vmatrix} h & k \\ h' & k' \end{vmatrix},$$

after reduction by subtracting the first row from each of the others.

In the (u, v)-plane we have a triangle $P_1' P_2' P_3'$, the coordinates of whose vertices are given by the relations

$$u_1 = f(x_1, y_1), \qquad\qquad v_1 = g(x_1, y_1),$$
$$u_2 = f(x_2, y_2) = f(x_1+h, y_1+k), \quad v_2 = g(x_2, y_2) = g(x_1+h, y_1+k),$$
$$u_3 = f(x_3, y_3) = f(x_1+h', y_1+k'), \quad v_3 = g(x_3, y_3) = g(x_1+h', y_1+k').$$

Expanding by Taylor's theorem, we obtain the *approximate* relations

$$u_2 = u_1 + h\frac{\partial f}{\partial x_1} + k\frac{\partial f}{\partial y_1}, \quad v_2 = v_1 + h\frac{\partial g}{\partial x_1} + k\frac{\partial g}{\partial y_1},$$

$$u_3 = u_1 + h'\frac{\partial f}{\partial x_1} + k'\frac{\partial f}{\partial y_1}, \quad v_3 = v_1 + h'\frac{\partial g}{\partial x_1} + k'\frac{\partial g}{\partial y_1}.$$

Hence, by a reduction similar to that adopted for A, the area of the triangle $P'Q'R'$ is A', where, approximately,

$$2A' = \pm \begin{vmatrix} h\dfrac{\partial f}{\partial x_1} + k\dfrac{\partial f}{\partial y_1}, & h\dfrac{\partial g}{\partial x_1} + k\dfrac{\partial g}{\partial y_1} \\[2ex] h'\dfrac{\partial f}{\partial x_1} + k'\dfrac{\partial f}{\partial y_1}, & h'\dfrac{\partial g}{\partial x_1} + k'\dfrac{\partial g}{\partial y_1} \end{vmatrix},$$

and so, using the well-known theorem on the product of two determinants (or verifying by direct calculation), we have the formula

$$2A' = \pm \begin{vmatrix} h & k \\ h' & k' \end{vmatrix} \times \begin{vmatrix} \dfrac{\partial f}{\partial x_1} & \dfrac{\partial g}{\partial x_1} \\ \dfrac{\partial f}{\partial y_1} & \dfrac{\partial g}{\partial y_1} \end{vmatrix}$$

$$= \pm 2A \frac{\partial(f,g)}{\partial(x_1, y_1)}.$$

Hence
$$A' = \pm A \frac{\partial(f,g)}{\partial(x_1, y_1)}.$$

Remembering that
$$u_1 \equiv f(x_1, y_1), \quad v_1 \equiv g(x_1, y_1),$$

we obtain the formula for a small area in the vicinity of a point (x_1, y_1), in the form

$$A' = \pm A \frac{\partial(u_1, v_1)}{\partial(x_1, y_1)},$$

the sign being adjusted to make the right-hand side positive.

7. Ratio of volumes under transformation. (Approximate theory.)

The formula of the preceding paragraph may be extended readily to three variables. Suppose that the relations

$$u = f(x, y, z), \quad v = g(x, y, z), \quad w = h(x, y, z)$$

define u, v, w as single-valued functions of the variables x, y, z. Take (x, y, z) as coordinates in one three-dimensional space and (u, v, w) as coordinates in another. We prove that, *if V is the volume of a small figure round the point (x_1, y_1, z_1), and if V' is the volume of the corresponding small figure round (u_1, v_1, w_1), then, approximately,*

$$V' = \pm \frac{\partial(u_1, v_1, w_1)}{\partial(x_1, y_1, z_1)} V,$$

the sign being adjusted to make the right-hand side positive.

Breaking the volume into small tetrahedra analogous to the triangles of § 6, we consider a particular one, of vertices (x_1, y_1, z_1), (x_2, y_2, z_2), (x_3, y_3, z_3), (x_4, y_4, z_4), where

$$x_2 = x_1 + p, \quad y_2 = y_1 + q, \quad z_2 = z_1 + r,$$

$$x_3 = x_1 + p', \quad y_3 = y_1 + q', \quad z_3 = z_1 + r',$$

$$x_4 = x_1 + p'', \quad y_4 = y_1 + q'', \quad z_4 = z_1 + r''.$$

The volume V is given by the formula

$$6V = \pm \begin{vmatrix} x_1 & y_1 & z_1 & 1 \\ x_2 & y_2 & z_2 & 1 \\ x_3 & y_3 & z_3 & 1 \\ x_4 & y_4 & z_4 & 1 \end{vmatrix}$$

$$= \pm \begin{vmatrix} p & q & r \\ p' & q' & r' \\ p'' & q'' & r'' \end{vmatrix},$$

on subtracting the first row from each of the others in turn.

After transformation we obtain a tetrahedron of vertices (u_i, v_i, w_i), where $u_1 = f(x_1\ y_1, z_1)$, $v_1 = g(x_1, y_1, z_1)$, $w_1 = h(x_1, y_1, z_1)$, $u_2 = f(x_1+p, y_1+q, z_1+r)$, $v_2 = g(x_1+p, y_1+q, z_1+r)$,
$$w_2 = h(x_1+p, y_1+q, z_1+r),$$
etc., and so, using Taylor's theorem, we have the approximate relations

$$u_2 = u_1 + p\frac{\partial f}{\partial x_1} + q\frac{\partial f}{\partial y_1} + r\frac{\partial f}{\partial z_1},$$

$$v_2 = v_1 + p\frac{\partial g}{\partial x_1} + q\frac{\partial g}{\partial y_1} + r\frac{\partial g}{\partial z_1},$$

$$w_2 = w_1 + p\frac{\partial h}{\partial x_1} + q\frac{\partial h}{\partial y_1} + r\frac{\partial h}{\partial z_1}, \quad \text{etc.}$$

Hence, approximately, the value of $6V'$ is

$$\pm \begin{vmatrix} p\dfrac{\partial f}{\partial x_1} + q\dfrac{\partial f}{\partial y_1} + r\dfrac{\partial f}{\partial z_1}, & p\dfrac{\partial g}{\partial x_1} + q\dfrac{\partial g}{\partial y_1} + r\dfrac{\partial g}{\partial z_1}, & p\dfrac{\partial h}{\partial x_1} + q\dfrac{\partial h}{\partial y_1} + r\dfrac{\partial h}{\partial z_1} \\ p'\dfrac{\partial f}{\partial x_1} + q'\dfrac{\partial f}{\partial y_1} + r'\dfrac{\partial f}{\partial z_1}, & p'\dfrac{\partial g}{\partial x_1} + q'\dfrac{\partial g}{\partial y_1} + r'\dfrac{\partial g}{\partial z_1}, & p'\dfrac{\partial h}{\partial x_1} + q'\dfrac{\partial h}{\partial y_1} + r'\dfrac{\partial h}{\partial z_1} \\ p''\dfrac{\partial f}{\partial x_1} + q''\dfrac{\partial f}{\partial y_1} + r''\dfrac{\partial f}{\partial z_1}, & p''\dfrac{\partial g}{\partial x_1} + q''\dfrac{\partial g}{\partial y_1} + r''\dfrac{\partial g}{\partial z_1}, & p''\dfrac{\partial h}{\partial x_1} + q''\dfrac{\partial h}{\partial y_1} + r''\dfrac{\partial h}{\partial z_1} \end{vmatrix}$$

$$= \pm \begin{vmatrix} p & q & r \\ p' & q' & r' \\ p'' & q'' & r'' \end{vmatrix} \times \begin{vmatrix} \dfrac{\partial f}{\partial x_1} & \dfrac{\partial g}{\partial x_1} & \dfrac{\partial h}{\partial x_1} \\ \dfrac{\partial f}{\partial y_1} & \dfrac{\partial g}{\partial y_1} & \dfrac{\partial h}{\partial y_1} \\ \dfrac{\partial f}{\partial z_1} & \dfrac{\partial g}{\partial z_1} & \dfrac{\partial h}{\partial z_1} \end{vmatrix}$$

$$= \pm 6V \frac{\partial(f, g, h)}{\partial(x_1, y_1, z_1)},$$

so that $$V' = \pm V \frac{\partial(u_1, v_1, w_1)}{\partial(x_1, y_1, z_1)}.$$

Note. The formulae

$$A = \pm A \frac{\partial(u_1, v_1)}{\partial(x_1, y_1)}, \quad V = \pm V \frac{\partial(u_1, v_1, w_1)}{\partial(x_1, y_1, z_1)}$$

imply that the Jacobians

$$\frac{\partial(u_1, v_1)}{\partial(x_1, y_1)}, \quad \frac{\partial(u_1, v_1, w_1)}{\partial(x_1, y_1, z_1)}$$

do not vanish. We have seen in § 1 (p. 90), that this vanishing is associated with points where the 'inverse' of the given transformation is not determined uniquely; but we do not propose to go more fully into the phenomenon.

REVISION EXAMPLES XII

University Level

1. Evaluate $\dfrac{\partial(\lambda, \mu)}{\partial(x, y)}$ in terms of λ, μ, where

$$\frac{x^2}{a+\lambda} + \frac{y^2}{b+\lambda} = 1, \quad \frac{x^2}{a+\mu} + \frac{y^2}{b+\mu} = 1.$$

2. If a, b, c, d are constants such that $ad - bc \neq 0$, and if

$$z = x^3 + y^3, \quad x = a\xi + b\eta, \quad y = c\xi + d\eta,$$

show that
$$
\begin{vmatrix} \dfrac{\partial^2 z}{\partial x^2} & \dfrac{\partial^2 z}{\partial x\,\partial y} \\[2mm] \dfrac{\partial^2 z}{\partial y\,\partial x} & \dfrac{\partial^2 z}{\partial y^2} \end{vmatrix} = \begin{vmatrix} \dfrac{\partial^2 z}{\partial \xi^2} & \dfrac{\partial^2 z}{\partial \xi\,\partial \eta} \\[2mm] \dfrac{\partial^2 z}{\partial \eta\,\partial \xi} & \dfrac{\partial^2 z}{\partial \eta^2} \end{vmatrix} \left\{ \frac{\partial(\xi, \eta)}{\partial(x, y)} \right\}^2.
$$

3. Six related variables f, g, u, v, x, y are such that any two of u, v, x, y can be taken as independent variables and the others are differentiable functions of these two. The partial derivative of f with respect to x when y is constant is denoted by $(\partial f/\partial x)_y$, and so on, and

$$\frac{\partial(f, g)}{\partial(x, y)} \equiv \begin{vmatrix} \left(\dfrac{\partial f}{\partial x}\right)_y & \left(\dfrac{\partial f}{\partial y}\right)_x \\[3mm] \left(\dfrac{\partial g}{\partial x}\right)_y & \left(\dfrac{\partial g}{\partial y}\right)_x \end{vmatrix}.$$

State without proof the formula for $\left(\dfrac{\partial f}{\partial x}\right)_y$ in terms of $\left(\dfrac{\partial f}{\partial u}\right)_v$, $\left(\dfrac{\partial f}{\partial v}\right)_u$, $\left(\dfrac{\partial u}{\partial x}\right)_y$, $\left(\dfrac{\partial v}{\partial x}\right)_y$, and show that

$$\frac{\partial(f,g)}{\partial(x,y)} = \frac{\partial(f,g)}{\partial(u,v)} \frac{\partial(u,v)}{\partial(x,y)}.$$

Hence, or otherwise, show that

$$\left(\frac{\partial v}{\partial y}\right)_u \left(\frac{\partial u}{\partial x}\right)_y = \frac{\partial(u,v)}{\partial(x,y)}$$

and

$$\left(\frac{\partial v}{\partial y}\right)_u = -\left(\frac{\partial u}{\partial y}\right)_v \left(\frac{\partial v}{\partial u}\right)_y.$$

4. Show that, if u, v are functions of x, y such that x, y are expressible uniquely as functions of u, v, then

$$\frac{\partial u}{\partial x}\frac{\partial x}{\partial u} + \frac{\partial v}{\partial x}\frac{\partial x}{\partial v} = 1, \quad \frac{\partial u}{\partial x}\frac{\partial y}{\partial u} + \frac{\partial v}{\partial x}\frac{\partial y}{\partial v} = 0,$$

with other similar equations.

If

$$J \equiv \frac{\partial(u,v)}{\partial(x,y)} \equiv \begin{vmatrix} \dfrac{\partial u}{\partial x} & \dfrac{\partial u}{\partial y} \\ \dfrac{\partial v}{\partial x} & \dfrac{\partial v}{\partial y} \end{vmatrix},$$

prove that

$$\frac{\partial(J,v)}{\partial(x,y)} = J\frac{\partial J}{\partial u}$$

and that

$$\frac{\partial}{\partial u}\left(\frac{1}{J}\frac{\partial u}{\partial x}\right) + \frac{\partial}{\partial v}\left(\frac{1}{J}\frac{\partial v}{\partial x}\right) = 0.$$

5. When v is eliminated between the equations

$$y = f(x,v), \quad z = g(x,v),$$

the equation $z = \phi(x,y)$ is obtained. Prove that

$$\frac{\partial\phi}{\partial x}\frac{\partial f}{\partial v} = \frac{\partial f}{\partial v}\frac{\partial g}{\partial x} - \frac{\partial f}{\partial x}\frac{\partial g}{\partial v}.$$

Verify this result when

$$y = x\cos v - a\sin v,$$
$$z = x\sin v + a\cos v,$$

a being a constant.

6. Each of the variables x, y, z is a function of the variables u, v. A partial derivative such as $\dfrac{\partial x}{\partial u}$ is formed on the assumption

that v is kept constant. The symbol $\left(\dfrac{\partial x}{\partial u}\right)_y$ implies that x is expressed as a function of u, y and y is kept constant during differentiation. The symbol $\dfrac{\partial(x, y)}{\partial(u, v)}$ is defined by the relation

$$\frac{\partial(x, y)}{\partial(u, v)} \equiv \begin{vmatrix} \dfrac{\partial x}{\partial u} & \dfrac{\partial y}{\partial u} \\[2ex] \dfrac{\partial x}{\partial v} & \dfrac{\partial y}{\partial v} \end{vmatrix}.$$

Prove the relations $\left(\dfrac{\partial x}{\partial u}\right)_y = \dfrac{\partial(x, y)}{\partial(u, v)}\Big/\dfrac{\partial y}{\partial v}$,

$$\left(\frac{\partial y}{\partial x}\right)_z = \frac{\partial(y, z)}{\partial(u, v)}\Big/\frac{\partial(x, z)}{\partial(u, v)}.$$

7. If x, y are defined as functions of z by the equations

$$f(x, y, z) = 0, \quad g(x, y, z) = 0,$$

show that $\dfrac{dx}{dz} = \dfrac{\partial(f, g)}{\partial(y, z)}\Big/\dfrac{\partial(f, g)}{\partial(x, y)}$,

where

$$\frac{\partial(f, g)}{\partial(y, z)} \equiv \begin{vmatrix} \dfrac{\partial f}{\partial y} & \dfrac{\partial g}{\partial y} \\[2ex] \dfrac{\partial f}{\partial z} & \dfrac{\partial g}{\partial z} \end{vmatrix}.$$

Find $\dfrac{dx}{dz}$ in this way when the defining equations are

$$z^2 + 2z - x - y + 2 = 0,$$
$$xy + y^2 - x + y - 1 = 0.$$

8. If x, y, z are functions of u, v with continuous first-order partial derivatives, show that

$$\frac{\partial(y, z)}{\partial(u, v)}\, dx + \frac{\partial(z, x)}{\partial(u, v)}\, dy + \frac{\partial(x, y)}{\partial(u, v)}\, dz = 0,$$

where

$$\frac{\partial(y, z)}{\partial(u, v)} \equiv \begin{vmatrix} \dfrac{\partial y}{\partial u} & \dfrac{\partial z}{\partial u} \\[2ex] \dfrac{\partial y}{\partial v} & \dfrac{\partial z}{\partial v} \end{vmatrix}.$$

Find $\dfrac{\partial z}{\partial x}$ in terms of u, v if

$$x = u^3 + v, \quad y = u + v^3, \quad z = e^{u^2 + v^2}.$$

CHAPTER XVII

MULTIPLE INTEGRALS

The ideas already explained for the integrals of functions of a single variable may now be extended to 'multiple integrals' of functions of several. The aim of this chapter is to give a clear picture of what is involved, but it must be recognized that the treatment is introduc tory. Once the general processes are understood, a detailed investigation may be found in a text-book of analysis; until the processes *are* understood, the necessity for the details can hardly be grasped.

We begin by the extension to functions of TWO variables.

1. Double integrals. Let
$$f(x,y)$$
be a function of the two variables x, y, defined for values of x, y in the region R of the (x, y)-plane (fig. 114) bounded by a simple closed curve C (without 'crossings').

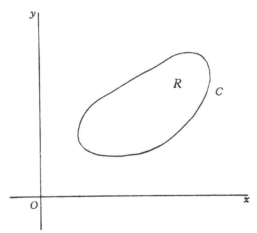

Fig. 114.

Divide the region R, in any manner, into a large number of small elements, of which a typical member has area ω_{ij}, and let M_{ij}, m_{ij} be the greatest and least values of $f(x, y)$ in ω_{ij}. Form the two sums
$$S_{ij} \equiv \Sigma M_{ij}\omega_{ij},$$
$$s_{ij} \equiv \Sigma m_{ij}\omega_{ij},$$

summed for all the elements ω_{ij}. We call S_{ij}, s_{ij} the *upper and lower sums* respectively for this particular subdivision.

Now suppose, analogously to integration for a single variable, that the number of elements ω_{ij} is increased indefinitely, the size of each being diminished indefinitely. In favourable cases, S_{ij} tends to a limiting value S which is independent of the method of subdivision, and s_{ij} tends similarly to a limiting value s.

The function $f(x,y)$ is said to be *integrable over the region R* if

$$S = s.$$

The common limit is then called the DOUBLE INTEGRAL of $f(x,y)$ over R.

In what follows we shall assume the integrability without further discussion, restricting our choice of functions to that end.

The actual evaluation, which is our main purpose, is made simpler by noticing that, if (ξ_i, η_j) is *any* point in the element of area ω_{ij}, then, by definition of M_{ij}, m_{ij}

$$M_{ij} \geqslant f(\xi_i, \eta_j) \geqslant m_{ij},$$

so that $\qquad \Sigma M_{ij}\omega_{ij} \geqslant \Sigma f(\xi_i, \eta_j)\,\omega_{ij} \geqslant \Sigma m_{ij}\omega_{ij}.$

On proceeding to the limit, the two outside sums assume the same value, to which
$$\Sigma f(\xi_i, \eta_j)\,\omega_{ij}$$

must therefore tend also. Hence *the greatest and least values M_{ij}, m_{ij} may be replaced in the definition by the value of $f(x,y)$ at any point in the element of area ω_{ij}*.

ILLUSTRATION 1. *To find the moment of inertia of a uniform rectangular lamina, of sides $2a$, $2b$ and density ρ, about a line through the centre parallel to the sides of length $2b$.*

Take axes, as shown in the diagram (fig. 115), through the centre and parallel to the sides. Divide the rectangle into a large number of small elements of which a typical member has area ω_{ij}, and let $P(x_i, y_j)$ be a point inside ω_{ij}. The moment of inertia about the axis Oy is defined to be the limit (if it exists) of the sum

$$\rho\Sigma x_i^2 \omega_{ij}$$

taken over all the elements as their number increases indefinitely while the size of each decreases indefinitely.

We have said nothing so far about the method of subdivision, but obviously an orderly system must be used if the calculation is to be at all manageable. The obvious way here is to divide by lines parallel to the axes (fig. 116). We therefore begin the calculation by considering the elements in a certain strip parallel to Oy (denoted by darker lines in the diagram), later completing the summation by adding the results obtained in all such strips.

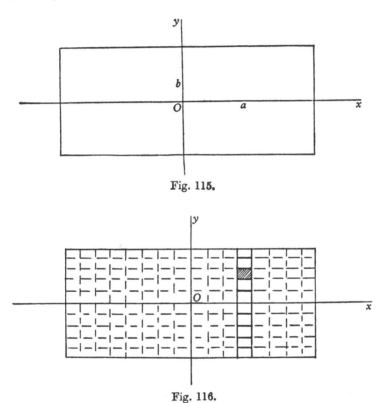

Fig. 115.

Fig. 116.

Suppose, then, that the strip parallel to Oy has a *fixed* width which we may call δx_i, and that a typical element in it has height δy_j; then $\omega_{ij} = \delta x_i \delta y_j$. Suppose, too, that a typical point inside this element has x-coordinate x_i, taken to be the same for each element of the strip. The required sum is

$$\rho \sum_i \sum_j x_i^2 \, \delta x_i \, \delta y_j.$$

To sum the elements in the strip, we make the summation with respect to j, keeping x_i and δx_i constant. By the definition of simple integration with respect to the variable y, this summation leads in the limit to the expression

$$\rho \sum_i x_i^2 \, \delta x_i \int_{-b}^{b} dy$$

$$= \rho \sum_i x_i^2 \, \delta x_i \left[\, y \,\right]_{-b}^{b}$$

$$= 2b\rho \sum_i x_i^2 \, \delta x_i.$$

To sum over all the strips, we now let δx_i become indefinitely small and proceed to the limit. This gives

$$2b\rho \int_{-a}^{a} x^2 \, dx$$

$$= 2b\rho \left[\, \tfrac{1}{3} x^3 \,\right]_{-a}^{a}$$

$$= \tfrac{4}{3} a^3 b \rho.$$

If M is the mass of the lamina, so that $M = 4\rho ab$, then the moment of inertia about Oy is
$$\tfrac{1}{3} M a^2.$$

ILLUSTRATION 2. *To find the moment of inertia of a uniform lamina of density ρ, bounded by the ellipse*

$$\frac{x^2}{a^2} + \frac{y^2}{b^2} = 1,$$

about the axis Oy.

The moment of inertia is, by definition,

$$\rho \Sigma x_i^2 \omega_{ij},$$

as before (p. 110).

For the method of subdivision, we again proceed by drawing lines parallel to the axes, and then summing 'up' a typical strip between the lines
$$x = x_i, \quad x = x_i + \delta x_i,$$

as indicated in the diagram (fig. 117). The line $x = x_i$ meets the ellipse in two points whose coordinates are

$$\left\{x_i, \; -b \Big/ \left(1 - \frac{x_i^2}{a^2}\right)\right\}, \quad \left\{x_i, \; +b \Big/ \left(1 - \frac{x_i^2}{a^2}\right)\right\}.$$

The strip is divided into a number of rectangular elements by the lines $y = $ const., save that there are 'end-effects' at the curved boundary. It is easy to arrange the subdivision in such a way that there is only *one* irregular element at each end of the strip—for example, the arc of the ellipse may be divided into a number of equal sections and the lines parallel to the axes drawn through the points of division.

The sum $\rho \Sigma x_i^2 \omega_{ij}$

may now be considered under two headings:

(i) the sum for all those elements ω_{ij} which are complete rectangles;

(ii) the sum for the remaining elements, at the boundary of the ellipse.

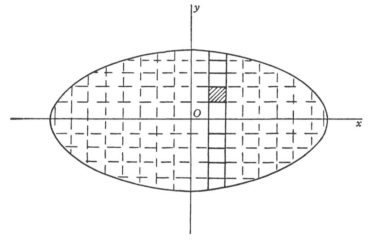

Fig. 117.

We first prove (what the beginner may prefer to regard as obvious) that the sum (ii) tends to zero as the subdivision approaches its limit.

The value of x_i^2 is certainly less than a^2, and the area ω_{ij} of an element is certainly less than its 'height' δy_j times the length δs_i of the element of arc bounding it (fig. 118). Thus the sum is less than

$$\Sigma \rho a^2 \delta s_i \delta y_j.$$

Hence, if η (which will tend to zero) is the greatest 'height' of any rectangle in the whole subdivision, so that $\delta y_j \leqslant \eta$, the sum is less than

$$\rho a^2 \eta \Sigma \delta s_i,$$

or $\rho a^2 \eta P,$

where P is the perimeter of the ellipse. In the limit, as η tends to zero, this sum also tends to zero, and so the sum (ii) is zero in the limit.

Since the 'end-effect' is zero, we may neglect it at once, and regard the strip of elements as running from precisely the value $-b\sqrt{\left(1-\dfrac{x_i^2}{a^2}\right)}$ to $+b\sqrt{\left(1-\dfrac{x_i^2}{a^2}\right)}$. Summing up the strip, we then have the sum

$$\rho\Sigma x_i^2\,\delta x_i\int_{-b\sqrt{(1-x_i^2/a^2)}}^{+b\sqrt{(1-x_i^2/a^2)}}dy$$

$$= 2\rho\Sigma bx_i^2\sqrt{\left(1-\frac{x_i^2}{a^2}\right)}\,\delta x_i.$$

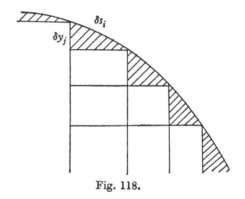

Fig. 118.

Summing now for all the strips, we obtain the limit

$$2\rho b\int_{-a}^{a}x^2\sqrt{\left(1-\frac{x^2}{a^2}\right)}dx.$$

If $x = a\sin\theta$, this is

$$2\rho ba^3\int_{-\frac{1}{2}\pi}^{\frac{1}{2}\pi}\sin^2\theta\cos\theta\cos\theta\,d\theta$$

$$= \tfrac{1}{2}\rho a^3 b\int_{-\frac{1}{2}\pi}^{\frac{1}{2}\pi}\sin^2 2\theta\,d\theta$$

$$= \tfrac{1}{4}\rho a^3 b\int_{-\frac{1}{2}\pi}^{\frac{1}{2}\pi}(1-\cos 4\theta)\,d\theta$$

$$= \tfrac{1}{4}\pi\rho a^3 b.$$

Since the mass of the lamina is

$$M = \pi\rho ab,$$

the value of the moment of inertia is

$$\tfrac{1}{4}Ma^2.$$

2. Notation. The method just illustrated for evaluating double integrals carries within itself the suggestion for the notation to be adopted. For Cartesian coordinates, where the subdivision is by rectangles, the integral is the limit of the expression

$$\Sigma f(x_i, y_j)\,\delta x_i\,\delta y_j$$

(evaluating the function, for convenience, at the corner (x_i, y_j) of the rectangle), and is written in the form

$$\iint f(x, y)\,dx\,dy.$$

We have just proved that the value of

$$\iint x^2\,dx\,dy$$

over the rectangle $|x| \leqslant a$, $|y| \leqslant b$ is

$$\tfrac{4}{3}a^3b,$$

and that its value over the interior of the ellipse $b^2x^2 + a^2y^2 = a^2b^2$ is

$$\tfrac{1}{4}\pi a^3b.$$

The value of $\qquad\qquad \iint dx\,dy$

over the region R bounded by a simple closed curve is equal to the area of R.

The preceding illustrations show how, in practice, a double integral is usually evaluated by considering it as a *repeated integral* with respect to the two variables in succession (in either order).

3. Triple integrals. The work in space follows by natural extension from that in a plane. Let

$$f(x, y, z)$$

be a function of the three variables x, y, z, defined for values of x, y, z in the region R enclosed by a simple closed surface F.

Divide the region R, in any manner, into a large number of small elements, of which a typical member has volume τ_{ijk}, and let M_{ijk},

m_{ijk} be the greatest and least values of $f(x, y, z)$ in τ_{ijk}. Form the *upper and lower sums*

$$S_{ijk} \equiv \Sigma M_{ijk}\tau_{ijk},$$

$$s_{ijk} \equiv \Sigma m_{ijk}\tau_{ijk},$$

summed for all the elements τ_{ijk}. If S_{ijk} tends to a limiting value S and s_{ijk} to a limiting value s (independently of the method of subdivision) as the number of elements τ_{ijk} increases indefinitely, the size of each decreasing indefinitely, and if, further, $S = s$, then we say that the function $f(x, y, z)$ is *integrable over the region* R. The common limit is then called the TRIPLE INTEGRAL of $f(x, y, z)$ over R.

As in the case of double integrals, we may for actual evaluation take the value of the function at a point (ξ_i, η_j, ζ_k) inside τ_{ijk}, and proceed to calculate the limiting value of the sum

$$\Sigma f(\xi_i, \eta_j, \zeta_k)\,\tau_{ijk}.$$

When the coordinates are Cartesian, the method of subdivision is naturally into 'boxes' of volume $\delta x_i\,\delta y_j\,\delta z_k$, and the limiting value of the sum

$$\Sigma f(x_i, y_j, z_k)\,\delta x_i\,\delta y_j\,\delta z_k$$

is denoted by the symbol

$$\iiint f(x, y, z)\,dx\,dy\,dz.$$

The method of evaluation usually consists in considering the triple integral as a *repeated integral* with respect to the variables in succession, taken in whatever order appears likely to be most convenient.

ILLUSTRATION 3. *To find the volume enclosed by the ellipsoid whose equation is*

$$\frac{x^2}{a^2} + \frac{y^2}{b^2} + \frac{z^2}{c^2} = 1.$$

Divide the ellipsoid into small 'boxes' by planes parallel to the axes of coordinates. A typical box, situated at the point (x_i, y_j, z_k) has volume $\delta x_i\,\delta y_j\,\delta z_k$ (with small corrections, which are negligible for sufficiently fine subdivision, at the surface itself*) and the total volume is the limit of the sum

$$\Sigma \delta x_i\,\delta y_j\,\delta z_k.$$

To begin with, keep x_i, y_j fixed, and integrate with respect to z up the 'tube' of cross-section δx_i, δy_j. This gives, for the volume of the tube, the formula

$$\Sigma \delta x_i\,\delta y_j \int dz,$$

---

* Compare the more detailed treatment on p. 113 for an analogous problem.

between appropriate limits, or

$$\Sigma \delta x_i \, \delta y_j [z].$$

Now, when $x = x_i$, $y = y_j$, the value of z varies along the tube from (approximately)

$$-c \sqrt{\left(1 - \frac{x_i^2}{a^2} - \frac{y_j^2}{b^2}\right)}$$

to

$$+c \sqrt{\left(1 - \frac{x_i^2}{a^2} - \frac{y_j^2}{b^2}\right)},$$

and so the volume of the tube is (approximately)

$$2c\Sigma \sqrt{\left(1 - \frac{x_i^2}{a^2} - \frac{y_j^2}{b^2}\right)} \delta x_i \, \delta y_j.$$

The totality of these tubes may be located by their sections in the (x, y)-plane $z = 0$, which they cut in small rectangles, covering the ellipse

$$\frac{x^2}{a^2} + \frac{y^2}{b^2} = 1$$

in that plane. In order to add all the tubes, we first keep, say, x_i constant and sum for all values of y within the ellipse, giving

$$2c\Sigma \delta x_i \int \sqrt{\left(1 - \frac{x_i^2}{a^2} - \frac{y^2}{b^2}\right)} dy$$

between appropriate limits of integration, namely,

$$y = -b \sqrt{\left(1 - \frac{x_i^2}{a^2}\right)}, \quad y = +b \sqrt{\left(1 - \frac{x_i^2}{a^2}\right)}.$$

To effect this integration, write

$$b \sqrt{\left(1 - \frac{x_i^2}{a^2}\right)} = \alpha,$$

so that the integral is

$$\frac{1}{b} \int_{-\alpha}^{\alpha} \sqrt{(\alpha^2 - y^2)} \, dy;$$

the substitution

$$y = \alpha \sin \theta$$

then gives

$$\frac{1}{b} \int_{-\frac{1}{2}\pi}^{\frac{1}{2}\pi} \alpha \cos \theta \, . \, \alpha \cos \theta \, d\theta$$

$$= \frac{\alpha^2}{b} \frac{\pi}{2}.$$

Hence the volume is $2c\Sigma\delta x_i \dfrac{\pi}{2b} \times b^2\left(1 - \dfrac{x_i^2}{a^2}\right),$

or, in the limit,

$$\pi bc \int_{-a}^{a} \left(1 - \frac{x^2}{a^2}\right) dx = \pi bc \left\{2a - \frac{2}{3}\frac{a^3}{a^2}\right\}$$

$$= \tfrac{4}{3}\pi abc.$$

EXAMPLES I

[It is most important that the reader should, at this stage, develop the faculty of, as it were, following the integration mentally up the filaments and then seeing the filaments move to cover the whole figure. The Illustrations in §4 show how the calculations are effected in practice, discarding the language of summation and proceeding straight to integration; the examples now given should be written out in detail like the above models until the picture of the process is completely clear. The later work should then be easy to follow.]

1. Find the moment of inertia, about the y-axis, of a uniform lamina of density ρ bounded by the straight lines

$$y = 1 + x, \quad y = 1 - x, \quad y = 0.$$

2. Find the mass of a lamina, lying in the positive quadrant, bounded by the axes and the circle $x^2 + y^2 = 1$, given that the density at any point is $k(1 + x^2)$. [Consider $\Sigma k(1 + x_i^2)\,\delta x_i\,\delta y_j$.]

3. Follow the method given in Illustration 3 (p. 116) to prove that the volume of a sphere of radius a is $\tfrac{4}{3}\pi a^3$.

4. Find the mass of a sphere of unit radius, given that the density at distance r from the centre is $k(1 + r^2)$.

[Consider $\Sigma k(1 + x_i^2 + y_j^2 + z_k^2)\,\delta x_i\,\delta y_j\,\delta z_k$.]

4. The evaluation in practice of multiple integrals. We now give some typical examples to show how multiple integrals are evaluated in normal practice. The *picture* is that of summation of elements; the *language* is that of definite integration.

(i) *To evaluate* $I \equiv \displaystyle\iint xy\,dx\,dy$

over the area in the first quadrant bounded by the straight line $x = 1$, the circle $x^2 + y^2 = 8$, and the parabola $y = \tfrac{1}{2}x^2$.

For a given value of x, the value of y 'up' the line $x = $ const. varies from $\frac{1}{2}x^2$ to $+\sqrt{(8-x^2)}$ (fig. 119). Hence

$$I = \int x\,dx\left[\tfrac{1}{2}y^2\right]_{\frac{1}{2}x^2}^{\sqrt{(8-x^2)}}$$

$$= \frac{1}{2}\int x\,dx(8 - x^2 - \tfrac{1}{4}x^4)$$

$$= \frac{1}{2}\int (8x - x^3 - \tfrac{1}{4}x^5)\,dx.$$

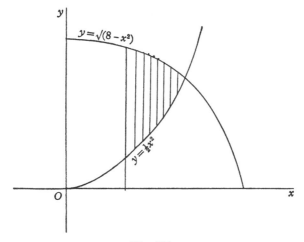

Fig. 119.

Now the range of values for x is (as the diagram indicates) from $x = 1$ to the value of x when the circle meets the parabola, namely, $x = 2$. Hence

$$I = \frac{1}{2}\int_1^2 (8x - x^3 - \tfrac{1}{4}x^5)\,dx$$

$$= \tfrac{1}{2}\left[4x^2 - \tfrac{1}{4}x^4 - \tfrac{1}{24}x^6\right]_1^2 = \tfrac{1}{2}[9\tfrac{1}{3} - 3\tfrac{17}{24}]$$

$$= 2\tfrac{13}{16}.$$

(ii) *To evaluate* $\qquad I \equiv \iint x\,e^{xy}\,dx\,dy$

over the rectangle bounded by the lines $x = 1$, $x = 2$, $y = 1$, $y = k$, *where* $k > 1$ (fig. 120).

Keeping x constant, first integrate with respect to y, from $y = 1$ to $y = k$. Since

$$\int_1^k x\,e^{xy}\,dy \quad (x \text{ constant})$$

$$= \left[e^{xy} \right]_1^k = (e^{kx} - e^x),$$

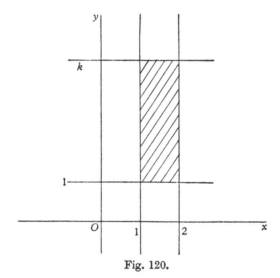

Fig. 120.

we have
$$I = \int_1^2 (e^{kx} - e^x)\,dx$$

$$= \left[\frac{1}{k} e^{kx} - e^x \right]_1^2$$

$$= \frac{1}{k} e^{2k} - \frac{1}{k} e^k - e^2 + e.$$

5. The double integral as a volume. We have seen (p. 13) how the function $f(x, y)$ is 'represented' by the surface whose equation is
$$z = f(x, y),$$
where z is the 'height' of the surface 'above' the point (x, y) in the plane $z = 0$. The definition of the double integral

$$\iint f(x, y)\,dx\,dy$$

by means of the sums $\Sigma M_{ij} \omega_{ij},\quad \Sigma m_{ij} \omega_{ij}$

shows at once that it may be used to evaluate the volume, 'above' the region R, contained between the surface $z = f(x, y)$ and the plane $z = 0$. (Compare the integral $\int_a^b f(x)\, dx$ as the area, 'above' the interval (a, b), contained between the curve $y = f(x)$ and the line $y = 0$.) In fact, $M_{ij}\omega_{ij}$ is the volume of a cylinder, standing on the area ω_{ij}, of height equal to or greater than $f(x_i, y_j)$; and $m_{ij}\omega_{ij}$ is the volume of a cylinder, standing on the area ω_{ij}, of height equal to or less than $f(x_i, y_j)$. The whole volume, obtained by summing these cylinders and proceeding to the limit, is thus

$$\iint f(x, y)\, dx\, dy.$$

ILLUSTRATION 4. *To find the volume of the 'box' whose base is the plane $z = 0$, whose sides are the planes $x = -1$, $x = 1$, $y = -2$, $y = 2$ and whose 'top' is the surface given by the equation*

$$x^2 + 5y^2 + z = 80.$$

The volume is $\qquad \iint (80 - x^2 - 5y^2)\, dx\, dy$

over the rectangle $x = \pm 1$, $y = \pm 2$.

Integrating first with respect to x, we have

$$\int \left[80x - \tfrac{1}{3}x^3 - 5y^2 x \right]_{x=-1}^{x=+1} dy$$

$$= \int_{-2}^{2} \{(80 - \tfrac{1}{3} - 5y^2) - (-80 + \tfrac{1}{3} + 5y^2)\}\, dy$$

$$= \int_{-2}^{2} (159\tfrac{1}{3} - 10y^2)\, dy$$

$$= \left[\tfrac{478}{3}y - \tfrac{10}{3}y^3 \right]_{-2}^{2}$$

$$= (\tfrac{956}{3} - \tfrac{80}{3}) - (-\tfrac{956}{3} + \tfrac{80}{3})$$

$$= 584.$$

6. 'Elements of area.' Double integrals.

In the definition of a double integral as a limit of summation, we dealt with the argument in terms of Cartesian coordinates, so that the element of area ω_{ij} was naturally chosen as a rectangle $\delta x_i\, \delta y_j$. The next problem is to consider what form of element should be taken when polar coordinates are used instead.

9

Suppose that the function to be integrated, over a region R bounded by a simple closed curve C, is expressed in terms of polar coordinates in the form

$$f(r, \theta).$$

Dividing the region into elements of area ω_{ij} and taking the value of the function at a point r_i, θ_j inside ω_{ij}, we reach as before the summation

$$\Sigma f(r_i, \theta_j)\, \omega_{ij}.$$

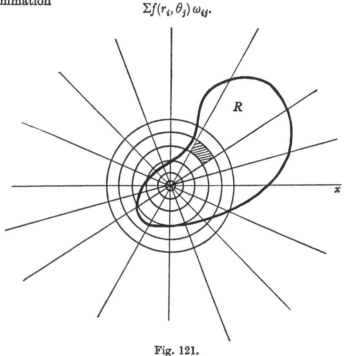

Fig. 121.

The value of the integral is the limit of this sum as the number of elements is increased indefinitely, the size of each decreasing indefinitely; it is assumed, as usual, that this limit is independent of the method of subdivision and of proceeding to the limit. What we have to do now is to find the most convenient shape for the element ω_{ij} and then to obtain a corresponding expression for its area.

The curves r constant are concentric circles whose centres are at the pole O, and the curves θ constant are straight lines through O. We are therefore led to consider an element, such as that shaded in the diagram (fig. 121), bounded by arcs of circles of radii r_i, $r_i + \delta r_i$

and by segments of straight lines inclined at angles θ_j, $\theta_j + \delta\theta_j$ to the initial line Ox. Subdivision into elements of this type will obviously be convenient for work in polar coordinates, provided that the expression for the area assumes a reasonably simple form.

Now the area of a sector (fig. 122) of angle $\delta\theta_j$ and radius $r_i + \delta r_i$ is known to be
$$\tfrac{1}{2}(r_i + \delta r_i)^2 \,\delta\theta_j.$$

(It bears the ratio $\delta\theta_j : 2\pi$ to the area of a complete circle of radius $r_i + \delta r_i$.) Hence the area of the element, being the difference of two sectors, of radii $r_i + \delta r_i$, r_i respectively, is
$$\tfrac{1}{2}\{(r_i + \delta r_i)^2 - r_i^2\}\,\delta\theta_j = r_i\,\delta r_i\,\delta\theta_j + \tfrac{1}{2}(\delta r_i)^2\,\delta\theta_j.$$

The summation required for the integration is thus
$$\Sigma f(r_i, \theta_j)\,r_i\,\delta r_i\,\delta\theta_j$$
$$+ \tfrac{1}{2}\Sigma f(r_i, \theta_j)\,(\delta r_i)^2\,\delta\theta_j,$$

where we have temporarily used the 'corner' (r_i, θ_j) of the element ω_{ij} as the point at which the function is evaluated.

Consider the second summation
$$U \equiv \tfrac{1}{2}\Sigma f(r_i, \theta_j)\,(\delta r_i)^2\,\delta\theta_j.$$

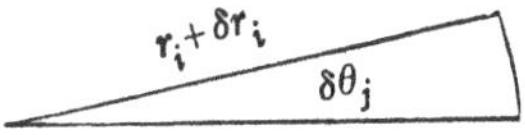

Fig. 122.

We prove* that it tends to zero with δr_i, $\delta\theta_j$.

It is assumed that the given function is bounded throughout the region R (fig. 121); suppose that it is numerically less than a certain constant K. Then,
$$U < \tfrac{1}{2}K\Sigma(\delta r_i)^2\,\delta\theta_j,$$
numerically.

Suppose next that, when the subdivision is made, all the elements δr_i are less than a certain number ρ, which itself is to shrink to zero for the limit. Then, numerically,
$$U < \tfrac{1}{2}K\rho\Sigma\delta r_i\,\delta\theta_j,$$

where we have succeeded in making the summation *linear* in δr_i.

Suppose further that the region R is bounded in extent, so that all points of it lie within a certain distance D of the pole O (fig. 123). For summation along a fixed sector $\delta\theta_j$, the value of $\Sigma\delta r_i$ certainly cannot exceed $2D$, the diameter of a circle containing the whole region R. Hence
$$U < K\rho D\Sigma\delta\theta_j,$$
numerically.

---

* The proof of this point may be omitted at a first reading.

Finally, the whole region R is covered as the radius vector θ_j swings from $\theta = 0$ to $\theta = 2\pi$ (or less), so that, by summation of $\delta\theta_j$,

$$U < 2\pi K\rho D,$$

numerically.

Since K, D are bounded, the right-hand side tends to zero with ρ, and so U tends to zero also.

Returning to the summation

$$\Sigma f(r_i, \theta_j)\, r_i \delta r_i \delta\theta_j$$

$$+ \tfrac{1}{2}\Sigma f(r_i, \theta_j)\, (\delta r_i)^2 \,\delta\theta_j,$$

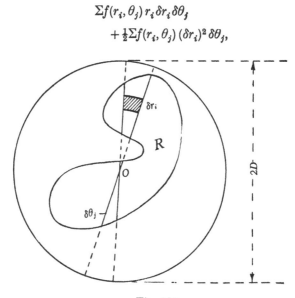

Fig. 123.

we are now able to ignore the second expression, since it vanishes in the limit. Hence the value of the integral is the limit of the sum

$$\Sigma f(r_i, \theta_j)\, r_i \delta r_i \delta\theta_j,$$

namely, $$\iint f(r, \theta)\, r\, dr\, d\theta.$$

This is therefore the formula for the integration of the function (r, θ) over the region R.

In actual practice, the formula just obtained is usually reached by directing the approximation to the element of area itself rather than to the second summation which we have just considered. Thus the elements are conceived as approximate rectangles of which two adjacent sides are of length $r_i \delta\theta_j$, δr_i, giving an approximate area

$$r_i \delta\theta_j \delta r_i$$

in agreement with our formula. As a *mnemonic* this treatment is useful, but it seems more satisfactory to trace the fate of the total 'correction term' and to see that it does really vanish.

At any rate, the formula for the double integral of a function $f(r, \theta)$ over a region R assumes a value which is the limit of the summation
$$\Sigma f(r_i, \theta_j) r_i \delta r_i \delta\theta_j,$$

written in the form
$$\iint f(r, \theta) r \, dr \, d\theta.$$

ILLUSTRATION 5. *To find the moment of inertia of a uniform lamina in the form of a cardioid $r = a(1 + \cos\theta)$, about a line through the origin perpendicular to its plane.*

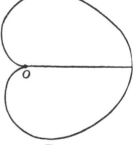

The moment of inertia is, by definition,

$$\rho \Sigma r_i^2 \delta\omega_{ij},$$

or, in the limit,

$$\rho \iint (r^2) r \, dr \, d\theta,$$

taken over the interior of the cardioid (fig. 124), where ρ is the density.

Fig. 124.

Keeping θ constant, integrate first with respect to r from $r = 0$ to $r = a(1 + \cos\theta)$. The result is

$$\rho \int d\theta [\tfrac{1}{4} r^4] = \tfrac{1}{4} a^4 \rho \int (1 + \cos\theta)^4 \, d\theta.$$

Now the whole cardioid is traced as θ moves from zero, through π, to 2π. Thus we have the expression

$$\tfrac{1}{4} a^4 \rho \int_0^{2\pi} (1 + 4\cos\theta + 6\cos^2\theta + 4\cos^3\theta + \cos^4\theta) \, d\theta,$$

where
$$\int_0^{2\pi} \cos\theta \, d\theta = \int_0^{2\pi} \cos^3\theta \, d\theta = 0,$$

$$\int_0^{2\pi} \cos^2\theta \, d\theta = \pi, \quad \int_0^{2\pi} \cos^4\theta \, d\theta = \tfrac{3}{4}\pi.$$

Hence the moment of inertia is

$$I = \tfrac{1}{4}a^4\rho(2\pi + 6\pi + \tfrac{3}{4}\pi) = \tfrac{35}{16}\pi a^4\rho.$$

Also the mass is, similarly,

$$M = \rho\iint r\,dr\,d\theta = \rho\int d\theta\,[\tfrac{1}{2}r^2]$$

$$= \tfrac{1}{2}a^2\rho\int_0^{2\pi} (1 + \cos\theta)^2\,d\theta = \tfrac{1}{2}a^2\rho(2\pi + \pi)$$

$$= \tfrac{3}{2}\pi a^2\rho.$$

Hence $$I = \tfrac{35}{24}Ma^2.$$

Note. It may be helpful to use this Illustration to clear up a
point which sometimes confuses the beginner. The sum

$$\rho\Sigma r_i^3\,\delta r_i\,\delta\theta_j$$

is really being interpreted in two ways:

(i) we derive it first in the form

$$\rho\Sigma r_i^2 . r_i\,\delta r_i\,\delta\theta_j$$

as the integral of the function ρr^2 for elements of area $r_i\,\delta r_i\,\delta\theta_j$ over
the interior of the cardioid $r = a(1 + \cos\theta)$;

(ii) we treat it next as the sum of terms ρr_i^3 for elements of area
$\delta r_i\,\delta\theta_j$, *as if the variables r_i, θ_j were* RECTANGULAR CARTESIAN
COORDINATES. (It was in this way, though not explicitly stated, that
we reconciled the sum $\rho\Sigma r_i^3\,\delta r_i\,\delta\theta_j$

with the formula $\rho\Sigma r_i^3\omega_{ij}$

leading to the integral $\rho\iint r^3\,dr\,d\theta.$)

The region of integration with θ, r as Cartesian coordinates is that
between $\theta = 0$ and $\theta = 2\pi$ lying 'under' the curve $r = a(1 + \cos\theta)$
—just like the area between $x = 0$ and $x = 2\pi$ lying under the
curve $y = a(1 + \cos x)$. This region is indicated in the diagram
(fig. 125).

In other words, we may say that the algebra involved in the
summation $\rho\Sigma r_i^3\,\delta r_i\,\delta\theta_j$

for $0 \leqslant r \leqslant a(1 + \cos\theta)$, $0 \leqslant \theta \leqslant 2\pi$ has two 'geometrical' interpretations:

(i) in polar coordinates r, θ, the integral of ρr^2 over the interior of the cardioid $r = a(1 + \cos\theta)$, in accordance with our original derivation;

(ii) in Cartesian coordinates θ, r (compare x, y), the integral of ρr^3 (compare ρy^3) over the area bounded by the straight lines $r = 0$, $\theta = 0$, $\theta = 2\pi$ and the 'cosine' curve $r = a(1 + \cos\theta)$. (Compare the area bounded by $y = 0$, $x = 0$, $x = 2\pi$ and the curve $y = a(1 + \cos x)$.)

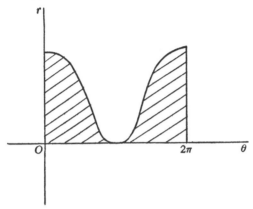

Fig. 125.

7. 'Elements of volume.' Triple integrals.

(i) CYLINDRICAL COORDINATES (p. 2). When a function to be integrated throughout a given volume is expressed in terms of cylindrical coordinates in the form

$$f(\rho, \phi, z),$$

the summation $\qquad \Sigma f(\rho_i, \phi_j, z_k)\, \tau_{ijk}$

is easily effected by taking the elements of volume as 'slices', $z = $ constant, standing on elements of area like those used (p. 122) in a plane for polar coordinates. By immediate extension we obtain the formula

$$\Sigma f(\rho_i, \phi_j, z_k)\, \rho_i\, \delta\rho_i\, \delta\phi_j\, \delta z_k,$$

leading to the limit $\qquad \displaystyle\iiint f(\rho, \phi, z)\, \rho\, d\rho\, d\phi\, dz.$

As an example, we obtain a formula which we shall use almost immediately afterwards. (Several other methods of calculation are available.)

ILLUSTRATION 6. *To prove that the volume of a 'sector' of a sphere of radius a bounded by a cone of vertical angle α, with its vertex at the centre of the sphere, is* $\frac{2}{3}\pi a^3(1-\cos\alpha).$

Take the axis of the cone as the z-axis and the centre O of the sphere as origin for a system of cylindrical coordinates ρ, ϕ, z. The volume is to be calculated from the formula

$$V = \iiint \rho\, d\rho\, d\phi\, dz.$$

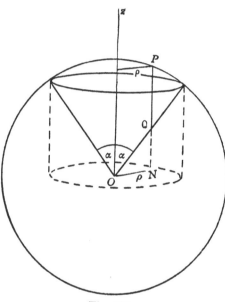

Fig. 126.

Keeping ρ, ϕ fixed integrate with respect to z up a 'filament' such as QP in the diagram (fig. 126). If this line meets the plane $z = 0$ in N, the length ON is ρ. Then, since P is on the sphere,

$$NP = (a^2-\rho^2)^{\frac{1}{2}},$$

and, from the triangle ONQ, in which $\angle OQN = \alpha$,

$$NQ = \rho\cot\alpha.$$

Hence the volume, being

$$\iint \rho\, d\rho\, d\phi\left[z\right]_Q^P,$$

is

$$\iint \rho\, d\rho\, d\phi\,\{(a^2-\rho^2)^{\frac{1}{2}}-\rho\cot\alpha\}.$$

The double integral is to be taken over the circle of radius $a \sin \alpha$ in the plane $z = 0$ (shown dotted in the diagram), which is the projection on that plane of the circle in which the cone meets the sphere. Keeping ρ fixed, integrate with respect to ϕ from 0 to 2π, giving

$$V = 2\pi \int_0^{a \sin \alpha} \rho \{ (a^2 - \rho^2)^{\frac{1}{2}} - \rho \cot \alpha \} \, d\rho.$$

By elementary integration, we have

$$V = 2\pi [-\tfrac{1}{3}(a^2 - \rho^2)^{\frac{3}{2}} - \tfrac{1}{3}\rho^3 \cot \alpha]_0^{a \sin \alpha}$$

$$= 2\pi a^3 \{ (-\tfrac{1}{3}\cos^3 \alpha - \tfrac{1}{3}\sin^3 \alpha \cot \alpha) + \tfrac{1}{3} \}$$

$$= \tfrac{2}{3}\pi a^3 (1 - \cos^3 \alpha - \sin^2 \alpha \cos \alpha)$$

$$= \tfrac{2}{3}\pi a^3 (1 - \cos \alpha).$$

(ii) SPHERICAL POLAR COORDINATES (p. 2). The evaluation of the sum

$$\Sigma f(r_i, \theta_j, \phi_k) \tau_{ijk}$$

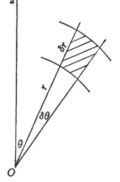

for spherical polar coordinates is more awkward. The surfaces r constant are spheres, the surfaces θ constant are cones, the surfaces ϕ constant are planes, and it is natural to use them to define the elements τ_{ijk}. We shall therefore proceed as follows.*

(a) Calculate the volume of the part of a sphere of radius r between two cones of vertical angles θ, $\theta + \delta\theta$.

(b) Calculate the difference of the volumes obtained in this way from two spheres of radii

Fig. 127.

r, $r + \delta r$; the figure may be visualized as the 'ring' obtained by rotating the area shaded in the diagram (fig. 127) about the axis Oz.

(c) Calculate the volume of the element of this ring contained between the planes ϕ, $\phi + \delta\phi$.

We take these calculations in turn.

(a) Using the formula obtained in the preceding Illustration 6, we have, for the volume between the two cones,

$$\tfrac{2}{3}\pi r^3 \{ [1 - \cos(\theta + \delta\theta)] - [1 - \cos \theta] \}$$

$$= \tfrac{2}{3}\pi r^3 \{ \cos \theta - \cos(\theta + \delta\theta) \}.$$

---

* But the timorous reader may proceed straight to the 'mnemonic' method given on p. 132.

But, by the mean-value theorem (Vol. I, p. 61), we have the relation

$$\cos(\theta + \delta\theta) = \cos\theta - \sin\theta'\,\delta\theta,$$

where θ' lies between θ and $\theta + \delta\theta$. Hence the volume is

$$\tfrac{2}{3}\pi r^3 \sin\theta'\,\delta\theta.$$

(b) Since θ, $\delta\theta$ are now being kept constant, the volume of the 'ring' contained between spheres of radii r, $r + \delta r$ is

$$\tfrac{2}{3}\pi\{(r + \delta r)^3 - r^3\}\sin\theta'\,\delta\theta$$

$$= 2\pi\{r^2\,\delta r + r(\delta r)^2 + \tfrac{1}{3}(\delta r)^3\}\sin\theta'\,\delta\theta.$$

(c) The fraction of the ring cut off between the two planes is $\delta\phi/2\pi$, and so, in all, we have

$$\tau_{ijk} = \{r^2\,\delta r + r(\delta r)^2 + \tfrac{1}{3}(\delta r)^3\}\sin\theta'\,\delta\theta\,\delta\phi.$$

Inserting suffixes, we may express the summation for the integral in the form

$$I + U + V,$$

where
$$I \equiv \Sigma f(r_i, \theta_j, \phi_k)\, r_i^2\,\delta r_i \sin\theta_j'\,\delta\theta_j\,\delta\phi_k,$$

$$U \equiv \Sigma f(r_i, \theta_j, \phi_k)\, r_i(\delta r_i)^2 \sin\theta_j'\,\delta\theta_j\,\delta\phi_k,$$

$$V \equiv \tfrac{1}{3}\Sigma f(r_i, \theta_j, \phi_k)\, (\delta r_i)^3 \sin\theta_j'\,\delta\theta_j\,\delta\phi_k.$$

The calculation, so far, is *exact*.

We prove that U (and similarly, by implication, V) tends to zero in the limiting process:

It is assumed that the given function is bounded throughout the volume of integration; suppose that it is numerically less than a certain constant K. Then

$$U < K\Sigma r_i(\delta r_i)^2 \sin\theta_j'\,\delta\theta_j\,\delta\phi_k$$

numerically.

Suppose next that, when the subdivision is made, all the elements δr_i are less than a certain number ρ, which itself is to shrink to zero for the limit. Then

$$U < K\rho\Sigma r_i\,\delta r_i \sin\theta_j'\,\delta\theta_j\,\delta\phi_k$$

numerically, where the summation now involves δr_i *linearly*.

Suppose further that the volume of integration is bounded in extent, so that all points of it lie within a certain distance D of the origin O. For summation along a fixed filament near the line at the position θ_j, ϕ_k, the value of $\Sigma r_i\,\delta r_i$ is certainly less than $D\Sigma\delta r_i$ numerically, or, at the worst, $2D^2$—since the sum $\Sigma\delta r_i$ along the filament cannot exceed the length $2D$ of the diameter of the sphere containing the whole volume of integration. Hence

$$U < 2K\rho D^2 \Sigma \sin\theta_j' \delta\theta_j \delta\phi_k$$

numerically.

Moreover, $\sin\theta_j'$ is numerically not greater than unity. Hence

$$U < 2K\rho D^2 \Sigma \delta\theta_j \delta\phi_k$$

numerically.

Keeping $\delta\phi_k$ constant, the sum $\Sigma\delta\theta_j$ cannot exceed π, as the radius vector θ constant swings from $\theta = 0$ to $\theta = \pi$. Thus

$$U < 2\pi K\rho D^2 \Sigma \delta\phi_k.$$

Finally, the whole volume is covered as the plane ϕ_k rotates from $\phi = 0$ to $\phi = 2\pi$ (or less), so that

$$U < 2\pi K\rho D^2 2\pi,$$
$$U < 4\pi^2 K\rho D^2$$

numerically.

Since K, D are bounded, the right-hand side tends to zero with ρ, so that U tends to zero.

It follows at once that V tends to zero also.

The summation giving the integral is therefore obtained from the expression

$$I = \sum_{(\text{limit})} f(r_i, \theta_j, \phi_k)\, r_i^2 \delta r_i \sin\theta_j' \delta\theta_j \delta\phi_k,$$

or, replacing θ_j by θ_j' in $f(r_i, \theta_j, \phi_k)$—as is legitimate since (p. 116) *any* point in the element of volume τ_{ijk} may be chosen at which to evaluate the function—we have

$$I = \sum_{(\text{limit})} f(r_i, \theta_j', \phi_k)\, r_i^2 \delta r_i \sin\theta_j' \delta\theta_j \delta\phi_k.$$

This limiting sum is, by definition, the integral denoted by the notation

$$I = \iiint f(r, \theta, \phi)\, r^2 \sin\theta\, dr\, d\theta\, d\phi.$$

The formula $r^2 \sin\theta\, dr\, d\theta\, d\phi$

for the 'element of volume' is usually recaptured by using as a mnemonic the figure shown in the diagram (fig. 128), which illustrates the surface of a sphere of radius r on which the two 'circles of latitude' $\theta, \theta + \delta\theta$ and the two 'circles of longitude' $\phi, \phi + \delta\phi$ are drawn. These delineate the shaded area, which may be regarded as a small 'rectangle' of sides $r\,\delta\theta$ (down the longitude) and $r\sin\theta\,\delta\phi$ (across the latitude). The element of area on the sphere is thus

$$r^2\sin\theta\,\delta\theta\,\delta\phi,$$

and multiplication by the 'height' δr gives the volume

$$r^2\sin\theta\,\delta r\,\delta\theta\,\delta\phi$$

of the element.

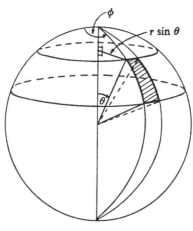

Fig. 128.

8. The elements by Jacobians; change of variables.

We proved in the preceding chapter (§§ 5, 6) that, for a transformation

$$u = f(x, y), \quad v = g(x, y)$$

between two planes in which x, y and u, v are taken *as Cartesian coordinates*, a small area A in the (x, y)-plane is transformed into a small area A' in the (u, v)-plane, where

$$A' = \pm A\,\frac{\partial(u, v)}{\partial(x, y)};$$

with the similar relation

$$V' = \pm V\,\frac{\partial(u, v, w)}{\partial(x, y, z)}$$

for volumes.

Taking first the case of a plane, consider the evaluation of the integral

$$\iint F(x, y)\,dx\,dy$$

over a region R. This is effected by means of the summation

$$\Sigma F(\xi_i, \eta_j)\, \omega_{ij}.$$

Under the transformation

$$u = f(x, y), \quad v = g(x, y),$$

the function $F(x, y)$ will become a certain function $G(u, v)$ of u, v; and evaluation at the point (ξ_i, η_j) will correspond to evaluation at a point (λ_i, μ_j) of the (u, v)-plane. The mesh of lines cutting out the elements ω_{ij} of the region R in the (x, y)-plane will give rise to a mesh of lines cutting out the elements Ω_{ij} of a region S in the (u, v)-plane. By definition, the integral of the function $G(u, v)$ over the region S is just $\Sigma G(\lambda_i, \mu_j)\, \Omega_{ij}$, *but this is* NOT *equal to the summation* $\Sigma F(\xi_i, \eta_j)\, \omega_{ij}$ *over* R. For, though each $G(\lambda_i, \mu_j)$ is equal to the corresponding $F(\xi_i, \eta_j)$, the element Ω_{ij} is the multiple $\pm \dfrac{\partial(u, v)}{\partial(x, y)}$ (evaluated at the point in question) of the corresponding element ω_{ij}. To make the *sums* equal, we must therefore reduce Ω_{ij} in the ratio $\pm 1 \Big/ \dfrac{\partial(u, v)}{\partial(x, y)}$, or (p. 98) $\pm \dfrac{\partial(x, y)}{\partial(u, v)}$. Thus

$$\Sigma F(\xi_i, \eta_j)\, \omega_{ij} = \Sigma G(\lambda_i, \mu_j) \left| \frac{\partial(\xi_i, \eta_j)}{\partial(\lambda_i, \mu_j)} \right| \Omega_{ij},$$

so that, in the limit, the relation for the double integrals is

$$\iint_R F(x, y)\, dx\, dy = \iint_S G(u, v) \left| \frac{\partial(x, y)}{\partial(u, v)} \right| du\, dv.$$

Finally, although we have given this explanation in terms of rectangular Cartesian coordinates, the summations themselves are independent of their geometrical interpretation, provided that the correct range of the variables u, v is taken. Compare the explanation given for the Illustration 5 on p. 126.

In particular, if u, v are the polar coordinates r, θ given by the relation

$$x = r \cos\theta, \quad y = r \sin\theta,$$

then

$$\frac{\partial(x, y)}{\partial(r, \theta)} = \begin{vmatrix} \cos\theta & -r\sin\theta \\ \sin\theta & r\cos\theta \end{vmatrix} = r,$$

so that the formula of transformation is

$$\iint F(x, y)\, dx\, dy = \iint G(r, \theta)\, r\, dr\, d\theta$$

over appropriate ranges of values for r, θ.

(Compare p. 124 for the 'element of area', $r\,dr\,d\theta$.)

Similar considerations apply to volumes. In particular, for cylindrical coordinates ρ, ϕ, z, where

$$x = \rho\cos\phi, \quad y = \rho\sin\phi, \quad z = z,$$

we have

$$\frac{\partial(x, y, z)}{\partial(\rho, \phi, z)} = \begin{vmatrix} \cos\phi & -\rho\sin\phi & 0 \\ \sin\phi & \rho\cos\phi & 0 \\ 0 & 0 & 1 \end{vmatrix} = \rho,$$

and the formula of transformation is

$$\iiint F(x,y,z)\,dx\,dy\,dz = \iiint G(\rho, \phi, z)\,\rho\,d\rho\,d\phi\,dz;$$

and for spherical polar coordinates r, θ, ϕ, where

$$x = r\sin\theta\cos\phi, \quad y = r\sin\theta\sin\phi, \quad z = r\cos\theta,$$

we have

$$\frac{\partial(x, y, z)}{\partial(r, \theta, \phi)} = \begin{vmatrix} \sin\theta\cos\phi & r\cos\theta\cos\phi & -r\sin\theta\sin\phi \\ \sin\theta\sin\phi & r\cos\theta\sin\phi & r\sin\theta\cos\phi \\ \cos\theta & -r\sin\theta & 0 \end{vmatrix}$$

$$= r^2\sin\theta \begin{vmatrix} \sin\theta\cos\phi & \cos\theta\cos\phi & -\sin\phi \\ \sin\theta\sin\phi & \cos\theta\sin\phi & \cos\phi \\ \cos\theta & -\sin\theta & 0 \end{vmatrix}$$

$$= r^2\sin\theta\{\cos\theta\cos\theta + \sin\theta\sin\theta\}$$

on expanding in terms of the last row. Hence

$$\frac{\partial(x, y, z)}{\partial(r, \theta, \phi)} = r^2\sin\theta,$$

and

$$\iiint F(x,y,z)\,dx\,dy\,dz = \iiint G(r, \theta, \phi)\,r^2\sin\theta\,dr\,d\theta\,d\phi.$$

(Compare p. 131 for the 'element of volume', $r^2\sin\theta\,dr\,d\theta\,d\phi$.)

The formulae

$$\iint F(x,y)\,dx\,dy = \iint G(u,v)\left|\frac{\partial(x,y)}{\partial(u,v)}\right|\,du\,dv,$$

$$\iiint F(x,y,z)\,dx\,dy\,dz = \iiint G(u,v,w)\left|\frac{\partial(x,y,z)}{\partial(u,v,w)}\right|\,du\,dv\,dw$$

enable us to make a transformation from variables x, y, z to new variables u, v, w. They are analogous to the formula

$$\int F(x)\, dx = \int F(u)\frac{dx}{du}\, du$$

for a single variable.

The Illustration which follows shows how such transformations may be used to simplify the evaluation of multiple integrals.

ILLUSTRATION 7. *To show that the triple integral*

$$\iiint \frac{1}{z^3}\sqrt{\left(\frac{x^2}{a^2}+\frac{y^2}{b^2}+\frac{z^2}{c^2}\right)}\, dx\, dy\, dz$$

taken throughout the volume common to the ellipsoid $\dfrac{x^2}{a^2}+\dfrac{y^2}{b^2}+\dfrac{z^2}{c^2}=1$

and the part of the cone $\alpha^2 x^2+\beta^2 y^2 = z^2$, *for which z is positive, has the value* $\pi/\alpha\beta$.

We can convert the ellipsoid into a sphere by the substitution

$$x = au, \quad y = bv, \quad z = cw.$$

Since

$$\frac{\partial(x, y, z)}{\partial(u, v, w)} = \begin{vmatrix} a & 0 & 0 \\ 0 & b & 0 \\ 0 & 0 & c \end{vmatrix} = abc,$$

we require the integral

$$\iiint \frac{1}{c^3 w^3}\sqrt{(u^2+v^2+w^2)}\, abc\, du\, dv\, dw$$

throughout the volume common to the sphere $u^2+v^2+w^2 = 1$ and the part of the cone $\alpha^2 a^2 u^2 + \beta^2 b^2 v^2 = c^2 w^2$, for which w is positive.

Transform now to spherical polars by the substitution

$$u = r\sin\theta\cos\phi, \quad v = r\sin\theta\sin\phi, \quad w = r\cos\theta.$$

We have

$$\frac{ab}{c^2}\iiint \left(\frac{r^3\sin\theta}{r^3\cos^3\theta}\right) dr\, d\theta\, d\phi = \frac{ab}{c^2}\iiint \sec^2\theta\tan\theta\, dr\, d\theta\, d\phi.$$

Keeping θ, ϕ constant, integrate with respect to r from $r = 0$ to $r = 1$, giving

$$\frac{ab}{c^2}\iint \sec^2\theta\tan\theta\, d\theta\, d\phi.$$

Now for points on the bounding cone, we have

$$\alpha^2 a^2 \sin^2\theta\cos^2\phi + \beta^2 b^2 \sin^2\theta\sin^2\phi = c^2\cos^2\theta.$$

Keeping ϕ constant, integrate with respect to θ from $\theta = 0$ to $\theta = \lambda$, where

$$\tan^2 \lambda = \frac{c^2}{\alpha^2 a^2 \cos^2 \phi + \beta^2 b^2 \sin^2 \phi},$$

giving

$$\frac{ab}{2c^2} \int d\phi \left[\tan^2 \theta \right]_0^\lambda = \frac{ab}{2c^2} \int \tan^2 \lambda \, d\phi$$

$$= \frac{ab}{2c^2} \int \frac{c^2}{\alpha^2 a^2 \cos^2 \phi + \beta^2 b^2 \sin^2 \phi} \, d\phi.$$

The cone is traced out by the rotation of ϕ from 0 to 2π. Since (as will appear at once) we are to divide numerator and denominator by $\cos^2 \phi$ to complete the integration, we avoid crossing the 'infinite' values of $\tan \phi$ which will arise by using symmetry in the four quadrants and evaluating

$$4 \times \frac{ab}{2c^2} \int_0^{\frac{1}{2}\pi} \frac{c^2}{\alpha^2 a^2 \cos^2 \phi + \beta^2 b^2 \sin^2 \phi} \, d\phi,$$

or, putting $\tan \phi = t$,

$$2ab \int_0^\infty \frac{dt}{\alpha^2 a^2 + \beta^2 b^2 t^2}$$

$$= \frac{2a}{\beta^2 b} \int_0^\infty \frac{dt}{t^2 + \left(\dfrac{\alpha a}{\beta b} \right)^2}$$

$$= \frac{2a}{\beta^2 b} \frac{\beta b}{\alpha a} \left[\tan^{-1} \left(\frac{t\beta b}{\alpha a} \right) \right]_0^\infty$$

$$= \frac{2}{\alpha \beta} \left[\frac{\pi}{2} \right]$$

$$= \frac{\pi}{\alpha \beta}.$$

REVISION EXAMPLES XIII

University Level

1. Show that the polar equation of the circle of unit radius which passes through the origin and has the initial line as diameter is $r = 2 \cos \theta$.

Integrate the function $1 - (1/r)$ over the area of this circle.

2. A thin lamina covers the positive quadrant of the circle $x^2 + y^2 = a^2$ and has density ρxy at the point (x, y). Find its centre of mass.

3. Integrate $\sin(x+y)$ over the square $0 \leqslant x \leqslant \pi$, $0 \leqslant y \leqslant \pi$; and $x\,e^{xy}$ over the rectangle $0 \leqslant x \leqslant a$, $0 \leqslant y \leqslant b$.

4. Evaluate
$$\iint xy\,dx\,dy$$
extended over (i) the triangle of sides $x = 0$, $y = 0$, $x+y = 3a$; (ii) the triangle of sides $y = 0$, $y = x+a$, $x+y = 3a$; (iii) the quadrilateral of sides $x = 0$, $y = 0$, $y = x+a$, $x+y = 3a$ and vertices $(0,0)$, $(0,a)$, $(a,2a)$, $(3a,0)$.

5. Show that the value of
$$\iiint_V \frac{dx\,dy\,dz}{(xzy)^{\frac{1}{4}}}$$
is four times the volume in the (u, v, w) space which corresponds to V, where $x = u^2vw$, $y = uv^2w$, $z = uvw^2$.

6. Evaluate
$$\iint x^3y\,dx\,dy$$
over the positive quadrant of the ellipse $x^2/a^2 + y^2/b^2 = 1$.

7. Calculate the volume of the solid bounded by the surface $z = xy$ and the planes $x = 0$, $y = 0$, $x+y = 1$, $z = 0$.

8. Evaluate the double integral
$$\iint (x+y)^2\,dx\,dy$$
taken over the area bounded by the curves $xy = 1$, $x^2y = 1$, and the line $x = 2$.

9. A plate has a plane triangular base and its thickness at any point is proportional to the sum of the perpendicular distances of the point (measured in the plane of the base) from the three sides of the triangle. Prove that its mean thickness is equal to the thickness at the centroid of the triangle.

10. Evaluate the integral
$$\iint (x+y+a)\,dx\,dy$$
taken over the circular area $x^2 + y^2 \leqslant a^2$.

11. Evaluate the integral

$$\iint z \, dx \, dy,$$

where $\dfrac{z}{c} = \dfrac{x^2}{a^2} + \dfrac{y^2}{b^2}$ and the region of integration is the elliptic area $\dfrac{x^2}{a^2} + \dfrac{y^2}{b^2} \leqslant 1$.

12. If D is the interior of the closed curve whose equation, in ordinary polar coordinates, is $r = 2a(1 + \cos\theta)$, evaluate the double integral

$$\iint_D (x^2 + y^2)^{\frac{1}{4}} \, dx \, dy,$$

where $x = r\cos\theta$, $y = r\sin\theta$.

13. A bounded function $f(x, y)$ is defined in a certain simple closed region R of the x, y plane. Explain briefly what is meant by the double integral

$$I = \iint_R f(x, y) \, dx \, dy.$$

The variables x, y are expressed in terms of two new independent variables u, v so that $f(x, y) = F(u, v)$ and R becomes a region S of the u, v plane. State, without proof, the expression for I as a double integral over S.

If R is the region bounded by the four parabolas $y^2 = ax$, $y^2 = Ax$, $x^2 = by$, $x^2 = By$, where $0 < a < A$, $0 < b < B$, find new variables u, v such that S is a rectangle, and hence show that the area of R is

$$\tfrac{1}{3}(A - a)(B - b).$$

14. A solid sphere is bounded by the surface $x^2 + y^2 + z^2 = a^2$, and a cylindrical hole is drilled through the sphere, the boundary of the hole being part of the surface $x^2 + y^2 - bx = 0$, where $0 < b \leqslant a$. Express the volume V of material removed as a double integral taken over a suitable region of the xy-plane, and derive the formula

$$V = 4\iint (a^2 - r^2)^{\frac{1}{2}} r \, dr \, d\theta,$$

where this double integral is taken over the region defined by $0 \leqslant \theta \leqslant \tfrac{1}{2}\pi$, $0 \leqslant r \leqslant b\cos\theta$.

Find V when $b = a$.

15. Show that

$$\iint x^3 y f(x^2+y^2)\,dx\,dy = \frac{1}{8}\int_0^1 u^2 f(u)\,du,$$

where the double integral is taken over the area

$$x\geqslant 0,\quad y\geqslant 0,\quad x^2+y^2\leqslant 1.$$

Evaluate the integral

$$\iint xy(b^2x^2+a^2y^2)^{\frac{2}{3}}\,dx\,dy$$

over the area $x\geqslant 0$, $y\geqslant 0$, $b^2x^2+a^2y^2\leqslant a^2b^2$.

16. If x, y, z are the perpendicular distances of a point from the sides of an equilateral triangle of side $2a$, prove that the mean value, with respect to area, of xyz taken over the interior of the triangle is $\frac{1}{20}a^3\sqrt{3}$.

17. A uniform solid of density ρ is in the form of an anchor ring, generated by rotating a circle of radius a about a line q in its plane and distant l ($>a$) from the centre of the circle. Show that the moment of inertia of the ring about q is $2\pi^2\rho a^2 l(l^2+\frac{3}{4}a^2)$.

18. Evaluate the integral

$$\iiint (ax^2+by^2+cz^2)\,dx\,dy\,dz,$$

taken throughout the spherical volume $x^2+y^2+z^2\leqslant 1$.

19. Prove that
(i) the area of a plane sector extending from the origin to a curve $x=f(u)$, $y=\phi(u)$ is

$$\frac{1}{2}\int \begin{vmatrix} x & y \\ \dfrac{dx}{du} & \dfrac{dy}{du} \end{vmatrix} du;$$

(ii) the volume of a cone extending from the origin to a surface $x=f(u,v)$, $y=\phi(u,v)$, $z=\psi(u,v)$ is

$$\frac{1}{3}\iint \begin{vmatrix} x & y & z \\ \dfrac{\partial x}{\partial u} & \dfrac{\partial y}{\partial u} & \dfrac{\partial z}{\partial u} \\ \dfrac{\partial x}{\partial v} & \dfrac{\partial y}{\partial v} & \dfrac{\partial z}{\partial v} \end{vmatrix} du\,dv.$$

20. Express the integral

$$\int_0^{4a} dy \left\{ \int_{y^2/4a}^{y} f(x,y)\,dx \right\}$$

in polar coordinates.

Evaluate the integral when

$$f(x,y) \equiv \frac{x^2 - y^2}{x^2 + y^2}.$$

21. Give an account of triple integrals sufficient to enable you to find the moment of inertia of the ellipsoid

$$\frac{x^2}{a^2} + \frac{y^2}{b^2} + \frac{z^2}{c^2} = 1$$

about the axis $y = z = 0$.

22. Prove that the volume in the positive octant which lies inside the surface $y^2 + z^2 = 4ax$ and outside the surface $y^2 = 4k^2ax$ (where $k < 1$) and between the planes $x = 0$, $x = c$ is

$$ac^2\{\cos^{-1} k - k \sqrt{(1 - k^2)}\}.$$

23. If r_1, r_2 are the distances of the point (x, y) from the foci $(\pm c, 0)$ of the ellipse $x^2/a^2 + y^2/b^2 = 1$, show, by means of the transformation $x + iy = c \cosh(u + iv)$ or otherwise, that

$$\iint \left(\frac{1}{r_1} + \frac{1}{r_2} \right) dx\,dy = 4\pi b,$$

the integral being taken over the interior of the ellipse.

24. By transforming to polar coordinates, show that the integral of the function $(x^2 + y^2)^2/(xy)^2$, taken over the area common to the circles $x^2 + y^2 = ax$, $x^2 + y^2 = by$ (where $a > 0$, $b > 0$), is ab.

Verify the result by transforming to variables u, v given by the relations $ux = vy = x^2 + y^2$.

25. A uniform plane lamina, of density ρ per unit area, is formed by the area common to the ellipse $x^2/a^2 + y^2/b^2 = 1$ and the parabola $2ay^2 = 3b^2x$. Find the moment of inertia of the lamina about the axis $x = 0$.

26. Show that the value of

$$\iint e^{(y-x)/(y+x)}\,dx\,dy$$

over the triangle bounded by the three lines $x = 0$, $y = 0$, $x + y = 1$ is equal to $(e^2 - 1)/4e$.

27. Prove that the quadric $1 + z = a^2x^2 + b^2y^2$ meets the sphere $x^2 + y^2 + z^2 = 1$ in a curve the polar coordinates (r, θ, ϕ) of any point of which satisfy the equations

$$r = 1, \quad \cos\theta = \frac{(a^2 - 1) + (b^2 - 1)\tan^2\phi}{a^2 + b^2\tan^2\phi},$$

and prove that the volume common to the sphere and the cone joining the origin to this curve is $2\pi/3ab$. It is assumed that a, b are each greater than unity.

28. Prove that the moment of inertia of a uniform lamina in the form of a limaçon $r = a + b\cos\theta$ (where $a > b$) about a line through the origin perpendicular to its plane is

$$\frac{(8a^4 + 24a^2b^2 + 3b^4)\,M}{8(2a^2 + b^2)},$$

where M is the mass of the lamina.

29. Show that the volume common to the sphere $x^2 + y^2 + z^2 = r^2$ and the ellipsoid $2x^2\sin^2\alpha + 2y^2\cos^2\alpha + z^2 = r^2$, where $0 < \alpha < \frac{1}{4}\pi$, is

$$\tfrac{4}{3}r^3(\tfrac{1}{2}\pi + 2\alpha\operatorname{cosec}2\alpha).$$

CHAPTER XVIII

THE SKETCHING OF CURVES

It is useful to be able to indicate by a rough sketch the general shape of a curve of given equation

$$f(x, y) = 0,$$

and the aim of the present chapter is to show, by means of particular examples, how to set about the problem. Details of the theory must be sought in a text-book on plane curves.

We saw in Volume I (p. 59) what features are most helpful when the equation can be reduced to the simple form

$$y = g(x),$$

where $g(x)$ is a function of x only. In the more general cases, with which we are now concerned, symmetry, gradient, concavity and so on—especially, perhaps, symmetry—form valuable guides, but more detailed analysis becomes necessary.

The following examples illustrate curves which the reader should be able to draw freely with little trouble, and they will be used without further explanation when required later. They are all of the form $y = Ax^n$, where n is a rational number, and so can be sketched by methods already given.

EXAMPLES I

Sketch the curves given by the equations:

1. $y = x^2$. 2. $y = x^3$.
3. $y^2 = x^3$. 4. $y = x^4$.
5. $y^3 = x^4$. 6. $y = x^5$.
7. $y^2 = x^5$. 8. $y^3 = x^5$.
9. $y^4 = x^5$. 10. $y^5 = x^6$.
11. $y^3 = x^2$. 12. $y^5 = x^3$.

We confine our attention to curves for which the function $f(x, y)$ is a *polynomial* in x, y. Such a curve is said to be ALGEBRAIC.

1. Towards a general method. Though our interest is centred in curves whose equation $f(x,y) = 0$ cannot be solved for y in terms of x, the general method may be illustrated by an example in which, exceptionally, the solution *can* be effected. This particular curve has what is called a 'singular point' at the origin, and we shall have to show later how such points may be located more generally. Apart from this, our treatment follows a standard process involving three stages:

(i) examination of the curve near the origin (or, if necessary, at other singular points);

(ii) examination of the curve at 'distant' points;

(iii) use of symmetry, special points (for example, points where the curve crosses the axes), ideas of continuity, and so on, to complete the curve.

ILLUSTRATION 1. *To sketch the curve*

$$y^2 = x^2 + 4x^3.$$

The solution for y in terms of x is immediate, giving the two possibilities

$$y = +x(1+4x)^{\frac{1}{2}}, \quad y = -x(1+4x)^{\frac{1}{2}}.$$

The curve is symmetrical about the x-axis; the positive sign corresponds to points 'above' the axis and the negative to points 'below'. It will therefore be sufficient to consider the former, namely,

$$y = x(1+4x)^{\frac{1}{2}}.$$

(i) *Small values of x.* When x is small, an approximation to the value of y may be obtained by means of the binomial theorem. Thus

$$y \simeq x(1+2x+...)$$
$$\simeq x + 2x^2 + ...$$

The first approximation $\quad y \simeq x$

shows that the curve lies very close to the line $y = x$, which is a tangent at the origin. The next approximation

$$y \simeq x + 2x^2$$

then shows that the value of y for the curve near the origin *exceeds* the value of y for the tangent. Hence the curve lies *above* the tangent, as the diagram (fig. 129) indicates.

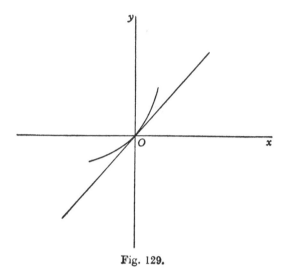

Fig. 129.

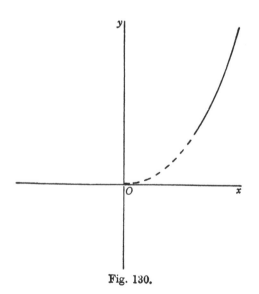

Fig. 130.

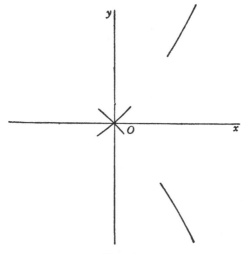

Fig. 131.

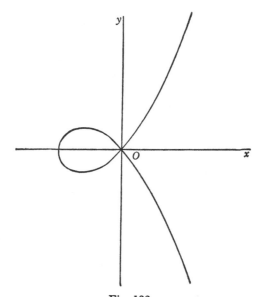

Fig. 132.

(ii) *Large values of x.* When x is large we again use the binomial theorem, but this time to express y in *descending* powers of x. Thus

$$y = 2x^{\frac{3}{2}}\left(1+\frac{1}{4x}\right)^{\frac{1}{2}}$$

$$\backsimeq 2x^{\frac{3}{2}}\left(1+\frac{1}{8x}+\ldots\right)$$

$$\backsimeq 2x^{\frac{3}{2}}+\tfrac{1}{4}x^{\frac{1}{2}}+\ldots.$$

The first approximation $y \backsimeq 2x^{\frac{3}{2}}$,

or $y \backsimeq 2x\sqrt{x}$,

shows that, for large values, x is necessarily positive, and the curve behaves very like the simpler curve $y^2 = 4x^3$ shown (for the 'upper' half) in the diagram (fig. 130).

(iii) *Intermediate values.* We have now, using symmetry, reached the stage indicated in fig. 131, and our problem is to 'join up' the arcs to complete the whole curve. The shape is made clear by noting that the curve meets the x-axis where $x^2(1+4x) = 0$, that is, in the origin and in the point $(-\tfrac{1}{4}, 0)$; and the y-axis where $y^2 = 0$, that is, in the origin only. We therefore obtain the form given in the diagram (fig. 132).

[It must be admitted that the final steps involve an appeal to a certain amount of intuition, and it is an intuition which the reader should develop. There is no other reasonable way of joining the arcs, given that the axes may not be crossed except at the point $(-\tfrac{1}{4}, 0)$.]

EXAMPLES II

Sketch the curves given by the equations:

1. $y^2 = x^2 - x^3$.
2. $y^2 = x^2 - 4x^4$.
3. $x^2 = y^2 + 4y^3$.
4. $x^2 = y^2 - 4y^4$.
5. $y^2 - 2xy + x^3 = 0$.
6. $y^2 + 4xy - 4x^4 = 0$.

2. The method for small values of x. We now seek a method* of 'successive approximation' to be used when simpler devices, such as the binomial theorem of the preceding illustration, are not

---

* We attempt to give a general method applicable to the many and varied complications. This one method, when correctly grasped, can be applied very widely, and seems to require a minimum amount of specialized information.

available. We give a number of examples to show the characteristic features which arise most frequently. In exposition, the first example will be given in detail, but after that less will be said in explanation as the process becomes more familiar.

The basic property, on which the method depends, will be seen by returning for a moment to the preceding Illustration 1 (i) (p. 143), where we obtained the successive approximations (for small x)

$$y \simeq x,$$

$$y \simeq x + 2x^2.$$

The first approximation is linear, corresponding to the tangent $y = x$; the next approximation is obtained by adding to the first a term of *higher* degree. This is what we seek to do in the examples which follow, where each approximation is obtained from its predecessor by bringing in further powers of x.

ILLUSTRATION 2. *To sketch the curve*

$$2x - y + x^2 + 3xy - y^3 = 0$$

for small values of x.

Near the origin the terms of lowest degree are most important, and we have the first approximation

$$2x - y \simeq 0$$

or $$y \simeq 2x.$$

The next approximation is therefore of the form

$$y \simeq 2x + Ax^n \quad (n > 1),$$

where A, n are constants to be determined, and where n is greater than 1 since the term Ax^n has to be small compared with $2x$. Substituting this approximation in the equation of the curve, we have the relation

$$-Ax^n + x^2 + 3x(2x + Ax^n) - (2x + Ax^n)^3 \simeq 0.$$

We are concerned with small values of x, so that terms of lowest degree are most important. In particular, in each of the expressions in brackets, namely, $2x + Ax^n$, where $n > 1$, the term Ax^n is negligible compared with $2x$ and may therefore be omitted; as a point of technique, we always imply such omissions by dots, thus:

$$-Ax^n + x^2 + 3x(2x + \ldots) - (2x + \ldots)^3 \simeq 0.$$

We then see that the final bracket, with terms in x^3, may be neglected entirely in the presence of earlier terms in x^2. We therefore have the position

$$-Ax^n + 7x^2 + \ldots \simeq 0.$$

All the other terms have been omitted by comparison with these; indeed, the comparison has been with the term $7x^2$, which is of lower degree than anything left out.

The essential point of the process now arises, namely, that, by choosing the particular values $A = 7$, $n = 2$, we reach a relation

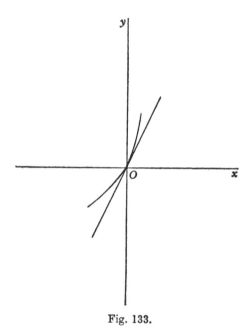

Fig. 133.

in which the left-hand side has no terms involving x to the power 2 or less, so that it approximates very closely to the (zero) value on the right. In other words, the values $A = 7$, $n = 2$ yield the approximation

$$y \simeq 2x + 7x^2$$

fitting the given equation of the curve to a high degree of accuracy.

It follows that the curve at the origin lies very close to the line $y = 2x$, which is the tangent there; also, since $7x^2$ is positive, the value of y for the curve exceeds the value of y for the tangent. Hence the curve lies 'above' the tangent (fig. 133).

ILLUSTRATION 3. (*Double point.*) *To sketch the curve*

$$2x^2 - 3xy + y^2 + x^2y + y^3 + x^4 = 0$$

for small values of x.

The terms of lowest degree give the approximation

$$2x^2 - 3xy + y^2 \simeq 0$$

or
$$(y - x)(y - 2x) \simeq 0.$$

There are therefore *two* possibilities: the curve may behave very like the straight line $y = x$, or very like the straight line $y = 2x$. The curve has a DOUBLE POINT (fig. 134) at the origin, with these lines as the tangents.

We take the two cases in turn:

(i) When $y \simeq x$,

the next approximation is

$$y \simeq x + Ax^n \quad (n > 1).$$

Experience shows that, when the terms of lowest degree factorize, the approximation should be inserted in the *factorized* form of equation. Thus, the equation is

$$(2x - y)(x - y) + x^2y + y^3 + x^4 = 0,$$

and the corresponding relation is

$$(x - Ax^n)(-Ax^n) + x^2(x + Ax^n) + (x + Ax^n)^3 + x^4 \simeq 0.$$

Omitting terms involving powers of x which (remembering that $n > 1$) are seen as negligible, we obtain the relation

$$(x - \ldots)(-Ax^n) + x^2(x + \ldots) + (x + \ldots)^3 + \ldots \simeq 0,$$

or
$$-Ax^{n+1} + 2x^3 + \ldots \simeq 0.$$

For closest approximation we cancel the terms of lowest degree, taking
$$A = 2, \quad n = 2.$$

Hence the second approximation to the value of y is

$$y \simeq x + 2x^2,$$

showing that the curve lies 'above' the tangent.

(ii) When $y \simeq 2x$,

the next approximation is

$$y \simeq 2x + Ax^n.$$

(It is convenient to use the same letters A, n, though, of course, they now have fresh values.)

Substituting and omitting obviously negligible terms, we have the relation

$$(-Ax^n)(-x-\dots)+x^2(2x+\dots)+(2x+\dots)^3+\dots \approx 0,$$

or $$Ax^{n+1}+10x^3+\dots \approx 0.$$

For closest approximation, we have

$$A=-10, \quad n=2,$$

giving $$y \approx 2x-10x^2,$$

so that the curve lies 'below' the tangent.

The shape of the curve near the origin in indicated in fig. 134.

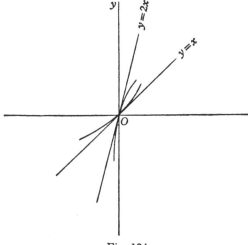

Fig. 134.

ILLUSTRATION 4. (*Cusp.*) *To sketch the curve*

$$4x^2-4xy+y^2+16x^2y+4y^3+x^4=0$$

for small values of x.

(The characteristic features of this equation are that the terms of lowest degree are quadratic, forming the square of a linear function which is not a factor of the cubic terms; that is, $2x-y$ is not a factor of $16x^2y+4y^3$.)

The terms of lowest degree give the approximation

$$4x^2-4xy+y^2 \approx 0$$

or $$(y-2x)^2 \approx 0,$$

so that $$y \approx 2x.$$

The next approximation is

$$y \simeq 2x + Ax^n \quad (n > 1).$$

Substituting and omitting obviously negligible terms, we have

$$A^2 x^{2n} + 16x^2(2x + \ldots) + 4(2x + \ldots)^3 + \ldots \simeq 0,$$

or

$$A^2 x^{2n} + 64x^3 + \ldots \simeq 0.$$

For closest approximation, we choose A, n so that

$$A^2 + 64 = 0, \quad 2n = 3,$$

or

$$A = \pm 8\sqrt{(-1)}, \quad n = \tfrac{3}{2}.$$

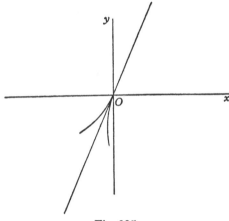

Fig. 135.

At first sight the presence of $\sqrt{(-1)}$ seems alarming, but it is essential to the argument. For the approximation is

$$y \simeq 2x \pm 8\sqrt{(-1)}\, x^{\frac{3}{2}},$$

or

$$y \simeq 2x \pm 8x\sqrt{(-x)}.$$

It follows that, for small values, x is necessarily negative; and also that the curve lies on either side of the tangent $y = 2x$, one 'branch' being 'above' it and the other 'below', as in the diagram (fig. 135).

At such a point, the curve is said to have a CUSP.

ILLUSTRATION 5. (*Tacnode.*) *To sketch the curve*

$$4x^2 - 4xy + y^2 + 2x^3 - x^2 y + 2xy^2 - y^3 - 12x^4 + y^4 = 0$$

for small values of x.

(The characteristic features of this equation are that the terms of lowest degree are quadratic, forming the square of a linear function which is also a factor of the cubic terms; that is, $2x-y$ is a factor of $2x^3 - x^2y + 2xy^2 - y^3 \equiv (2x-y)(x^2+y^2)$.)

Bearing in mind the remark on p. 149, we use the 'factorized' form of equation

$$(2x-y)^2 + (2x-y)(x^2+y^2) - 12x^4 + y^4 = 0.$$

As in the preceding example, we consider the approximation

$$y \simeq 2x + Ax^n \quad (n > 1).$$

Substituting and omitting obviously negligible terms, we have

$$A^2x^{2n} - Ax^n\{x^2 + (2x + \ldots)^2\} - 12x^4 + (2x + \ldots)^4 \simeq 0,$$

or $\qquad\qquad A^2x^{2n} - 5Ax^{n+2} + 4x^4 + \ldots \simeq 0.$

We have retained all terms which might conceivably be wanted; the others are seen at once to be negligible by comparison with their immediate neighbours.

It is now clear that the value $n = 2$ involves all three terms, so that we have, for closest approximation,

$$A^2 - 5A + 4 = 0, \quad n = 2.$$

Thus $\qquad\qquad A = 1 \text{ or } 4, \quad n = 2.$

We therefore have the two approximations

$$y \simeq 2x + x^2,$$
$$y \simeq 2x + 4x^2.$$

The curve thus consists of two touching branches, each lying 'above' the tangent on either side of the origin (fig. 136).

At such a point, the curve is said to have a TACNODE.

We conclude this paragraph with an example, not involving any essentially new principle, to demonstrate the slight modification necessary when the tangent at the origin is the y-axis:

ILLUSTRATION 6. *To sketch the curve*

$$x^2 + x^4 - y^4 + y^5 = 0$$

for small values of x.

The approximation $x^2 \simeq 0$ suggests that we interchange the roles of x, y and attempt to express x in increasing powers of y. We thus make the approximation

$$x \simeq Ay^n \quad (n > 1).$$

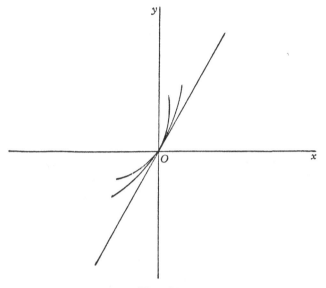

Fig. 136.

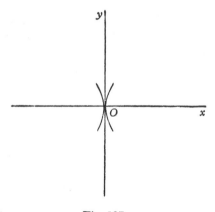

Fig. 137.

(Note that n is greater than 1, *not* 0; the first approximation was $x \doteq 0 \cdot y$.)

On substituting and omitting obviously negligible terms, we have

$$A^2 y^{2n} + \ldots - y^4 + \ldots \doteq 0,$$

or

$$A^2 y^{2n} - y^4 + \ldots \doteq 0.$$

For closest approximation, we choose A, n so that

$$A^2 = 1, \quad 2n = 4,$$

or $\quad\quad\quad\quad\quad A = \pm 1, \quad n = 2.$

Hence we have the two approximations

$$x \backsimeq y^2, \quad x \backsimeq -y^2,$$

leading to the form (a tacnode) shown in the diagram (fig. 137).

GENERALISATION. A point P on a curve given by an equation of degree k in x, y is called a MULTIPLE POINT, of multiplicity p, if an *arbitrary* line through P meets it in only $k - p$ points other than P.

It follows at once that, if P is the origin, the equation contains no terms of degree less than p and at least one term of degree p. (Compare Illustrations 3, 4, 5, where $k = 4$, $p = 2$.)

EXAMPLES III

Use the method of this section to obtain the approximations near the origin for the curves given in Examples II (p. 146).

3. The method for large values of x. Very similar methods may also be applied to study the form of a curve at large distances from the origin. Referring again to the preliminary Illustration 1 (ii) (p. 146), we recall that we were able there to express y in *decreasing* powers of x; we had the successive approximations

$$y \backsimeq 2x^{\frac{3}{2}}, \quad y \backsimeq 2x^{\frac{3}{2}} + \tfrac{1}{4}x^{\frac{1}{2}}.$$

In the more general case, we obtain our first approximation from the terms of *highest* degree in the equation $f(x, y) = 0$. The subsequent improvements are obtained just as in the case of small values of x. The Illustrations which follow exemplify a number of typical curves.

(*Note.* The calculations will of themselves justify the use of the terms of highest degree for the first approximation; when x, y are *both* large these terms have obvious importance. But it is possible for, say, x to remain finite while y becomes large, as in the familiar example of the rectangular hyperbola $xy = 1$. We hold this case back for some time (pp. 172–3).)

ILLUSTRATION 7. *To sketch the curve*

$$2x^3 - x^2y - 2xy^2 + y^3 - 2x^2 - 2xy + 2x + 8y = 0$$

for large values of x.

When x is very large, only the terms of highest degree are significant, so we have the first approximation

$$2x^3 - x^2y - 2xy^2 + y^3 \simeq 0$$

or
$$(x - y)(2x - y)(x + y) \simeq 0.$$

We therefore consider in turn three approximations to the value of y, given by $y \simeq x$, $y \simeq 2x$, $y \simeq -x$. For convenience, we shall complete the calculations for all three cases before commenting on the geometrical interpretations.

(i) For $y \simeq x$, the next approximation is

$$y \simeq x + Bx^p,$$

where p is *less* than 1. Substituting this value in the equation

$$(x - y)(2x - y)(x + y) - 2x^2 - 2xy + 2x + 8y = 0,$$

where the terms of highest degree are expressed in factorized form, we have
$$-Bx^p(x - Bx^p)(2x + Bx^p)$$
$$-2x^2 - 2x(x + Bx^p)$$
$$+2x + 8(x + Bx^p) \simeq 0.$$

We are now interested in the *larger* powers of x, so that we may omit terms of lesser degree. Remembering that $p < 1$, we thus have

$$-Bx^p(x - \ldots)(2x + \ldots) - 2x^2 - 2x(x + \ldots) + \ldots \simeq 0$$

or
$$-2Bx^{p+2} - 4x^2 + \ldots \simeq 0,$$

each omission (indicated by dots) being justified on comparison with neighbouring terms.

For closest approximation we choose B, p so that these terms vanish, giving
$$B = -2, \quad p = 0.$$

Hence the approximation is
$$y \simeq x - 2.$$

(ii) For $y \simeq 2x$, the next approximation is

$$y \simeq 2x + Bx^p \quad (p < 1)$$

(for fresh values of B, p), so that, on substituting and keeping terms of highest degree,

$$(-x - \ldots)(-Bx^p)(3x + \ldots) - 2x^2 - 2x(2x + \ldots) + \ldots \simeq 0$$

or
$$3Bx^{p+2} - 6x^2 + \ldots \simeq 0,$$

so that
$$B = 2, \quad p = 0.$$

Hence the approximation is
$$y \backsimeq 2x + 2.$$

(iii) For $y \backsimeq -x$, the next approximation is
$$y \backsimeq -x + Bx^p \quad (p < 1),$$
so that, as before,
$$(2x - \ldots)(3x - \ldots)(Bx^p) - 2x^2 - 2x(-x + \ldots) + \ldots \backsimeq 0$$
or
$$6Bx^{p+2} + 0 . x^2 + \ldots \backsimeq 0.$$

Strictly speaking, B is essentially non-zero, so we ought to con-
tinue the approximation by retaining further terms. But comparison with the two preceding approximations shows us that the
vanishing of the coefficient of x^2 is, in a sense, 'accidental'. We are
about to examine all three cases further, so, for uniformity of
treatment, we pause here and keep the approximation in the form
$$y \backsimeq -x.$$
We have therefore reached the three approximations
$$y \backsimeq x - 2,$$
$$y \backsimeq 2x + 2,$$
$$y \backsimeq -x.$$

Experience shows that the next approximations are best obtained by replacing the terms of highest degree $(x-y)(2x-y)(x+y)$
in the given equation by the corresponding product
$$(x - y - 2)(2x - y + 2)(x + y)$$
and making the consequential adjustments. Now we have, by
direct multiplication, the identity
$$(x - y - 2)(2x - y + 2)(x + y)$$
$$\equiv (x - y)(2x - y)(x + y) - 2x^2 - 2xy - 4x - 4y,$$
so that the equation of the curve is expressible in the form
$$(x - y - 2)(2x - y + 2)(x + y) + 6x + 12y = 0.$$

Consider the three approximations in turn:

(i) Let
$$y \backsimeq x - 2 + Cx^q,$$
where q is now less than 0. On substituting and keeping only terms
of highest degree, we have
$$(-Cx^q)(x + \ldots)(2x - \ldots) + 6x + 12(x - \ldots) \backsimeq 0,$$
or
$$-2Cx^{q+2} + 18x + \ldots \backsimeq 0,$$

so that, for closest approximation,

$$C = 9, \quad q = -1.$$

Hence

$$y \simeq x - 2 + \frac{9}{x}.$$

(ii) Let $\quad\quad y \simeq 2x + 2 + Cx^q \quad (q < 0).$

Then $\quad (-x - \ldots)(-Cx^q)(3x + \ldots) + 6x + 12(2x + \ldots) \simeq 0,$

or $\quad\quad\quad 3Cx^{q+2} + 30x + \ldots \simeq 0,$

so that $\quad\quad\quad C = -10, \quad q = -1.$

Hence

$$y \simeq 2x + 2 - \frac{10}{x}.$$

(iii) Let $\quad\quad y \simeq -x + Cx^q \quad (q < 0),$

where we know from the earlier work that q must be *less* than zero although, in this particular case, the constant term is absent. As before,

$$(2x - \ldots)(3x - \ldots)(Cx^q) + 6x + 12(-x + \ldots) \simeq 0,$$

or $\quad\quad\quad 6Cx^{q+2} - 6x + \ldots \simeq 0,$

so that $\quad\quad\quad C = 1, \quad q = -1.$

Hence

$$y \simeq -x + \frac{1}{x}.$$

To summarize, we have obtained the three approximations

$$y \simeq x - 2 + \frac{9}{x},$$

$$y \simeq 2x + 2 - \frac{10}{x},$$

$$y \simeq -x + \frac{1}{x}.$$

The geometrical significance may be appreciated most readily by referring to the sketch (not drawn to scale) in the diagram (fig. 138).

When x is large, y is very nearly equal to one or other of the expressions $x - 2$, $2x + 2$, $-x$, and so the curve (for large values of x) lies very near to the three straight lines

$$y = x - 2,$$

$$y = 2x + 2,$$

$$y = -x.$$

Straight lines which the curve approaches in this way are called ASYMPTOTES of the curve.

The further approximations then show that, if x is large and *positive*, the curve lies 'above' the asymptote $y = x - 2$, 'below' the asymptote $y = 2x + 2$, and 'above' the asymptote $y = -x$; if x is large and *negative*, the curve lies 'below' $y = x - 2$, 'above' $y = 2x + 2$, and 'below' $y = -x$. These features should be checked by reference to the diagram.

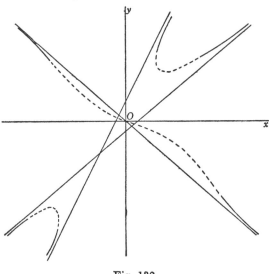

Fig. 138.

ILLUSTRATION 8. (*Parallel asymptotes.*) *To sketch the curve*

$$x^3 + x^2y - xy^2 - y^3 + 6xy + 6y^2 - 19x - 11y + 6 = 0$$

for large values of x.

The terms of highest degree give the approximation

$$x^3 + x^2y - xy^2 - y^3 \approx 0$$

or
$$(x - y)(x + y)^2 \approx 0.$$

We therefore have two approximations

$$y \approx x,$$

$$y \approx -x,$$

of which the second may be expected to present fresh features; we begin with the normal case $y \approx x$.

Taking the equation of the curve in the form

$$(x-y)(x+y)^2 + 6xy + 6y^2 - 19x - 11y + 6 = 0,$$

we make the approximation

$$y \simeq x + Bx^p \quad (p < 1),$$

so that

$$(-Bx^p)(2x + ...)^2 + 6x(x + ...) + 6(x + ...)^2 - ... \simeq 0,$$

or

$$-4Bx^{p+2} + 12x^2 + ... \simeq 0.$$

Hence

$$B = 3, \quad p = 0,$$

and the approximation is $\quad y \simeq x + 3.$

Suppose next that

$$y \simeq -x + Bx^p \quad (p < 1).$$

We notice that the terms of degree one less than the highest (that is, the quadratic terms) also contain $x + y$ as a factor—a point that should always be investigated when repeated factors occur—and so we write the given equation in the form

$$(x-y)(x+y)^2 + 6y(x+y) - 19x - 11y + 6 = 0.$$

Application of the routine process of substitution needs care. At first sight, we have

$$(2x - ...)(B^2x^{2p}) + 6(-x + ...)(Bx^p) - ... \simeq 0,$$

or

$$2B^2x^{2p+1} - 6Bx^{p+1} + ... \simeq 0,$$

so that $p = 0$. But a glance at the equation then shows that the terms $-19x - 11y$ are also involved, and we therefore incorporate them in the relation

$$(2x - ...)(B^2x^{2p}) + 6(-x + ...)(Bx^p) - 19x - 11(-x + ...) + ... \simeq 0,$$

where the terms now omitted are all negligible by comparison with their immediate neighbours. Thus

$$2B^2x^{2p+1} - 6Bx^{p+1} - 8x + ... \simeq 0,$$

and we take

$$p = 0,$$

$$2B^2 - 6B - 8 = 0,$$

or

$$B = -1, \quad +4,$$

giving the *two* approximations

$$y \simeq -x - 1,$$

$$y \simeq -x + 4.$$

We have obtained in all an asymptote

$$y = x + 3$$

and two parallel asymptotes

$$y = -x - 1,$$
$$y = -x + 4.$$

For the next approximation, we begin as before (p. 156), with the identity (obtained by direct multiplication of the three expressions derived from the approximations already obtained)

$$(y - x - 3)(y + x + 1)(y + x - 4)$$
$$\equiv (y - x)(y + x)^2 - 6xy - 6y^2 + 13x + 5y + 12,$$

so that the equation of the curve is

$$(y - x - 3)(y + x + 1)(y + x - 4) + 6(x + y) - 18 = 0.$$

(Note that we have again exhibited the factor $x + y$ in the term of highest remaining degree.)

The approximation

$$y \simeq x + 3 + Cx^q \quad (q < 0)$$

gives $Cx^q(2x + \ldots)(2x + \ldots) + 6(2x + \ldots) - \ldots \simeq 0,$

or $4Cx^{q+2} + 12x + \ldots \simeq 0,$

so that $C = -3, \quad q = -1,$

and the approximation is

$$y \simeq x + 3 - \frac{3}{x}.$$

The approximation

$$y \simeq -x - 1 + Cx^q \quad (q < 0)$$

gives $(-2x + \ldots)(Cx^q)(-5 + \ldots) + 6(-1 + \ldots) - 18 \simeq 0,$

or $10Cx^{q+1} - 24 + \ldots \simeq 0,$

so that $C = \frac{12}{5}, \quad q = -1,$

and the approximation is

$$y \simeq -x - 1 + \frac{12}{5x}.$$

The approximation

$$y \simeq -x + 4 + Cx^q \quad (q < 0)$$

gives $(-2x - \ldots)(5 + \ldots)(Cx^q) + 6(4 + \ldots) - 18 \simeq 0,$

or $-10Cx^{q+1}+6+\ldots \fallingdotseq 0,$

so that $C = \tfrac{3}{5}, \quad q = -1,$

and the approximation is

$$y \fallingdotseq -x+4+\frac{3}{5x}.$$

We have thus obtained three asymptotes, of which two are parallel; the further approximations show that (i) when x is large and positive the curve lies 'below' $y = x+3$ and 'above' $y = -x-1$ and $y = -x+4$; (ii) when x is large and negative the curve lies 'above' $y = x+3$ and 'below' $y = -x-1$ and $y = -x+4$. These

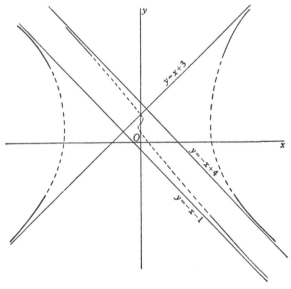

Fig. 139.

properties should be checked by reference to the diagram (fig. 139) which exhibits the main features of the curve but is not drawn to scale. (Note as 'guides' that the curve meets the y-axis in the points $(0, 1)$, $(0, 2)$, $(0, 3)$, that it crosses the asymptote $y = x+3$ at the point $(0, 3)$, and that it does not meet the parallel asymptotes.)

EXAMPLES IV

Use this method to find the asymptotes of the following curves:

1. $2x^2 - 3xy + y^2 + 3x - 2y - 1 = 0$.
2. $x^2 - y^2 + 2x - 2y + 1 = 0$.
3. $(x+y)(x-y)(2x-y) + 4x^2 - 2xy + 2y^2 - 2y + x + 7 = 0$.
4. $(x+y)(x-y)^2 + x^2 - xy - 3y = 0$.
5. $(x-y)(x+y)^2 + 4x^2 + 4xy + 1 = 0$.

ILLUSTRATION 9. (*Parabolic asymptotes; also, completing the sketch.*) *To sketch the curve*

$$x^3 - x^2 y - xy^2 + y^3 - x^2 + 3xy = 0.$$

The terms of highest degree are

$$(y+x)(y-x)^2,$$

so that, for large values of x, we have to consider the two approximations

$$y \simeq -x,$$

$$y \simeq x.$$

The first of these follows standard pattern, and we may deal with it quickly. The equation is

$$(y+x)(y-x)^2 - x^2 + 3xy = 0,$$

so that the approximation

$$y \simeq -x + Bx^p \quad (p < 1)$$

gives $\qquad (Bx^p)(-2x + ...)^2 - x^2 + 3x(-x + ...) \simeq 0,$

or $\qquad 4Bx^{p+2} - 4x^2 + ... \simeq 0,$

so that $\qquad p = 0, \quad B = 1,$

and there is therefore an asymptote

$$y = -x + 1.$$

The given equation, in the form

$$(y+x-1)(y-x)^2 + y(x+y) = 0,$$

leads from the approximation

$$y \simeq -x + 1 + Cx^q \quad (q < 0)$$

to the relation

$$(Cx^q)(-2x + ...)^2 + (-x + ...)(1 + ...) \simeq 0,$$

or
$$4Cx^{q+2} - x + \ldots \simeq 0,$$

so that we take
$$q = -1, \quad C = \tfrac{1}{4},$$

giving
$$y \simeq -x + 1 + \frac{1}{4x}.$$

Thus the curve lies 'above' the asymptote when x is positive and 'below' when x is negative.

The approximation
$$y \simeq x$$
introduces new features, so we treat it with more detail. Writing
$$y \simeq x + Bx^p,$$

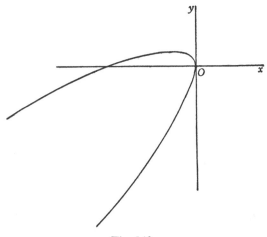

Fig. 140.

we have the relation
$$(2x + \ldots)(Bx^p)^2 - x^2 + 3x(x + \ldots) \simeq 0,$$

or
$$2B^2 x^{2p+1} + 2x^2 + \ldots \simeq 0,$$

so that
$$p = \tfrac{1}{2}, \quad B = \pm i \quad (i = \sqrt{(-1)}).$$

Thus
$$y \simeq x \pm \sqrt{(-x)},$$
which shows that x must be negative on this branch when numerically large.

The curve this time does *not* approach any straight line. On the other hand, it behaves like the two branches
$$y = x + \sqrt{(-x)}, \quad y = x - \sqrt{(-x)},$$
which, when considered together, appear as the two 'arms' of the parabola (fig. 140)
$$(y - x)^2 + x = 0.$$

Before commenting further, we proceed to the next approximation, writing the equation of the curve in the form

$$(y+x)\{(y-x)^2+x\} - 2x^2 + 2xy = 0,$$

or $\qquad (y+x)(y-x+i\sqrt{x})(y-x-i\sqrt{x}) + 2x(y-x) = 0.$

The approximation

$$y \simeq x + i\sqrt{x} + Cx^q \quad (q < \tfrac{1}{2})$$

gives $\qquad (2x + ...)(2i\sqrt{x} + ...)(Cx^q) + 2x(i\sqrt{x} + ...) \simeq 0$

or $\qquad 4iCx^{q+\frac{1}{2}} + 2ix^{\frac{3}{2}} + ... \simeq 0,$

so that $\qquad q = 0, \quad C = -\tfrac{1}{2}.$

Hence $\qquad y \simeq x + i\sqrt{x} - \tfrac{1}{2},$

with a similar result when i is replaced by $-i$. Thus a very close approximation to the curve for large values of x is found in the parabola
$$(y - x + \tfrac{1}{2})^2 + x = 0,$$

which is called the ASYMPTOTIC PARABOLA of the curve.

The earlier parabola, $(y-x)^2 + x = 0$, gives a good idea of the general appearance of the curve, and is usually found sufficient in practice. When more accurate 'placing' is required, the asymptotic parabola must be used.

This completes the investigation of the curve for large values of x. In order to sketch the whole curve, we now look at the small values. The terms of lowest degree give $x \simeq 0$ and $y \simeq \tfrac{1}{3}x$, so we examine each of them in turn.

For $x \simeq 0$, we proceed in ascending powers of y and consider (compare p. 152) the approximation

$$x \simeq Ay^n \quad (n > 1).$$

Substituting in the equation (arranged now in terms of ascending degree)
$$x(3y - x) + x^3 - x^2y - xy^2 + y^3 = 0,$$

we have $\qquad Ay^n(3y - ...) + ... + y^3 \simeq 0,$

retaining terms of lowest degree in y. Thus

$$3Ay^{n+1} + y^3 + ... \simeq 0,$$

so that, for best approximation, we have

$$n = 2, \quad A = -\tfrac{1}{3}.$$

Hence $\qquad x \simeq -\tfrac{1}{3}y^2,$

showing that, near the origin, the curve lies to the left of the tangent $x = 0$.

Similarly, if $\qquad y \simeq \frac{1}{3}x + Ax^n \quad (n > 1),$

then $\qquad x(3Ax^n) + x^3 - x^2(\frac{1}{3}x + \dots) - x(\frac{1}{3}x + \dots)^2 + (\frac{1}{3}x + \dots)^3 \simeq 0$

or $\qquad\qquad 3Ax^{n+1} + \frac{16}{27}x^3 + \dots \simeq 0,$

so that $\qquad\qquad n = 2, \quad A = -\frac{16}{81}.$

Thus $\qquad\qquad y \simeq \frac{1}{3}x - \frac{16}{81}x^2,$

showing that, near the origin, the curve lies 'below' the tangent $y = \frac{1}{3}x.$

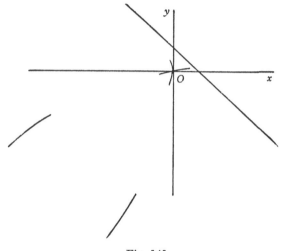

Fig. 141.

We have therefore reached the position given in the diagram (fig. 141).

For the rest of the curve we must revert to more general considerations. Unfortunately we receive no help from symmetry, but we are able to calculate the coordinates of the points where the curve crosses the axes and the linear asymptote.

When $x = 0$, we have $y = 0$ only; when $y = 0$, we have $x = 0, 0, 1$; when $x + y = 1$, we have $x = 1, y = 0$. Hence the curve does not meet the y-axis except at the origin, and it cuts the x-axis and the asymptote where they cross at the point $(1, 0)$. Thus we obtain the form in the diagram (fig. 142), which, however, is not drawn to scale.

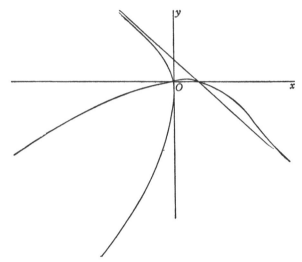

Fig. 142.

EXAMPLES V

Find the asymptotic parabola of the curves:

1. $2(2y-x)(y+x)^2 + 9xy = 0.$
2. $y^2 = 4x^2(x+y).$
3. $x(y+3x)^2 = y^2 + 9.$

ILLUSTRATION 10. (*Loop and inflexion.*) *To sketch the curve*

$$x^5 - y^5 + x^2(x+y) = 0.$$

For large values of x, we have

$$(x-y)(x^4 + x^3y + x^2y^2 + xy^3 + y^4) \fallingdotseq 0,$$

and the only real approximation is

$$y \fallingdotseq x.$$

(The presence of terms not resoluble into real linear factors may indicate the presence of a loop. But this is not necessarily the case.)

Writing $\qquad y \fallingdotseq x + Bx^p \quad (p < 1),$

we have $\qquad (-Bx^p)(5x^4 + ...) + x^2(2x + ...) \fallingdotseq 0,$

or $\qquad -5Bx^{p+4} + 2x^3 + ... \fallingdotseq 0,$

so that $\qquad B = \frac{2}{5}, \quad p = -1.$

We therefore have the approximation

$$y \eqsim x + \frac{2}{5x},$$

showing that the line $y = x$ is an asymptote, and also that the curve lies 'above' it when x is positive and 'below' it when x is negative.

When x is small, we have the two approximations $x^2 \eqsim 0$, $x + y \eqsim 0$, which we consider in turn.

If $\qquad\qquad x \eqsim Ay^n \quad (n > 1),$

then $\qquad\qquad -y^5 + A^2 y^{2n}(y + \ldots) \eqsim 0,$

or $\qquad\qquad -y^5 + A^2 y^{2n+1} + \ldots \eqsim 0,$

so that $\qquad\qquad n = 2, \quad A = \pm 1.$

We therefore have the two approximations

$$x \eqsim y^2,$$

$$x \eqsim -y^2,$$

indicating the presence of a tacnode (p. 153).

If $\qquad\qquad y \eqsim -x + Ax^n \quad (n > 1),$

then $\qquad\qquad x^5 - (-x + \ldots)^5 + x^2(Ax^n) \eqsim 0,$

or $\qquad\qquad 2x^5 + Ax^{n+2} + \ldots \eqsim 0,$

so that $\qquad\qquad n = 3, \quad A = -2.$

We therefore have the approximation

$$y \eqsim -x - 2x^3.$$

Thus the curve lies 'below' the tangent $y = -x$ when x is positive and 'above' when x is negative.

When the curve crosses the tangent in this way, the curve is said to have an INFLEXION at the point.

This particular curve has both a tacnode (tangent $x = 0$) and an inflexion (tangent $x + y = 0$) at the origin.

We have so far reached the state indicated in the diagram (fig. 143), and our problem is to combine these arcs into a single curve.

We note that, since replacing x, y by $-x$, $-y$ does not affect the equation, the curve is symmetrical about the origin. Also it meets the axes in the origin only, the asymptote $x = y$ in the origin only, and the line $x + y = 0$ in the origin only. Thus the curve does not cross any of the lines shown in the diagram, and so its form is that indicated in fig. 144. The arcs within the area bounded by the lines $x = 0$, $x + y = 0$ join up by means of loops.

4. To locate the multiple points of the curve $f(x, y) = 0$.
Suppose that the point (h, k), initially unknown, is a multiple point of the curve

$$f(x, y) = 0.$$

Referred to axes through that point parallel to the given axes, the equation assumes the form

$$f(x' + h, y' + k) = 0,$$

where $f(h, k)$ is, of course, zero. Expanding by Taylor's theorem (p. 57), we have

$$f(h, k) + \left(x' \frac{\partial}{\partial h} + y' \frac{\partial}{\partial k} \right) f(h, k) + \ldots = 0,$$

or
$$\frac{\partial f(h, k)}{\partial h} x' + \frac{\partial f(h, k)}{\partial k} y' + \frac{\partial^2 f(h, k)}{2! \, \partial h^2} x'^2 + \ldots = 0.$$

By hypothesis, the new origin is a multiple point of the curve and so the coefficients of x', y' both vanish. Thus

$$\frac{\partial f(h, k)}{\partial h} = 0, \quad \frac{\partial f(h, k)}{\partial k} = 0.$$

In other words, *the multiple points, if any, are those whose coordinates satisfy simultneously the three equations*

$$f(x, y) = 0,$$
$$\frac{\partial f(x, y)}{\partial x} = 0, \quad \frac{\partial f(x, y)}{\partial y} = 0.$$

ILLUSTRATION 11. *To find the multiple points of the curve*

$$y^3 + x^2 + 2xy - 6y^2 - 2x + 14y - 11 = 0.$$

The multiple points are those for which, on differentiating partially with respect to x, y in turn,

$$2x + 2y - 2 = 0,$$
$$3y^2 + 2x - 12y + 14 = 0.$$

Substituting for $2x$ in the second equation, we have

$$3y^2 - 14y + 16 = 0,$$

or
$$(y - 2)(3y - 8) = 0.$$

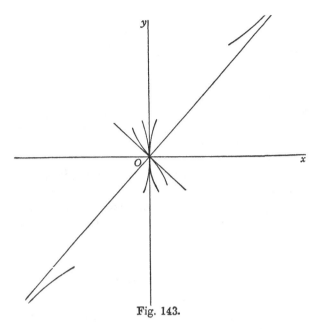

Fig. 143.

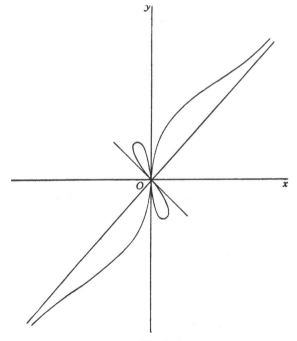

Fig. 144.

Thus $\qquad\qquad\qquad y = 2$ or $\frac{8}{3}$,

so that, from the equation $2x + 2y - 2 = 0$,

$$x = -1 \quad \text{or} \quad -\tfrac{5}{3}.$$

Thus possible multiple points are $(-1, 2)$, $(-\frac{5}{3}, \frac{8}{3})$. But it is easy to check that the solution $x = -1, y = 2$ satisfies the given equation whereas the solution $x = -\frac{5}{3}$, $y = \frac{8}{3}$ does not. Hence there is a multiple point at the point $(-1, 2)$.

5. Some rules for finding the asymptotes in simple cases.

The method which we have given enables us to find the asymptotes of a curve $f(x, y) = 0$ with reasonable ease, and also (what is often very important) to estimate how the curve lies in relation to them. There are, however, one or two rules available in simple cases.

The equation of the curve is

$$f(x, y) \equiv u_n(x, y) + u_{n-1}(x, y) + \ldots = 0,$$

where $u_k(x, y)$ is a polynomial homogeneous of degree k in x, y. We begin the process of finding the asymptotes by factorizing the polynomial $u_n(x, y)$ in the form

$$u_n(x, y) \equiv (y - k_1 x)(y - k_2 x) \ldots (y - k_n x)$$

and we restrict ourselves to the case when the constants $k_1, k_2, \ldots, k_n$ are *real and distinct*. With these restrictions, we can formulate the following rules:

(i) THE 'PARTIAL FRACTIONS' RULE FOR ASYMPTOTES. *If*

$$u_{n-1}(x, y)/u_n(x, y),$$

when expressed in partial fractions, assumes the form

$$\frac{u_{n-1}(x, y)}{u_n(x, y)} = \frac{\alpha_1}{y - k_1 x} + \frac{\alpha_2}{y - k_2 x} + \ldots + \frac{\alpha_n}{y - k_n x}$$

(where it is assumed also that $u_{n-1}(x, y)$, $u_n(x, y)$ do not have common factors), then the equations of the asymptotes are

$$y - k_i x + \alpha_i = 0 \quad (i = 1, 2, \ldots, n).$$

The given equation, after division by $u_n(x, y)$ is

$$1 + \frac{u_{n-1}(x, y)}{u_n(x, y)} + \frac{u_{n-2}(x, y)}{u_n(x, y)} + \ldots = 0,$$

or $\qquad 1 + \dfrac{\alpha_1}{y - k_1 x} + \ldots + \dfrac{\alpha_n}{y - k_n x} + \dfrac{u_{n-2}(x, y)}{u_n(x, y)} + \ldots = 0.$

Substitute the approximation

$$y \simeq k_1 x + B x^p \quad (p < 1)$$

in this form of the equation. Then

$$1 + \frac{\alpha_1}{Bx^p} + \frac{\alpha_2}{(k_1 - k_2)x + Bx^p} + \dots + \frac{\alpha_n}{(k_1 - k_n)x + Bx^p}$$
$$+ \frac{u_{n-2}(x, k_1 x + Bx^p)}{u_n(x, k_1 x + Bx^p)} + \dots \simeq 0.$$

Since $p < 1$, and since none of $k_1 - k_2, \dots, k_1 - k_n$ is zero, the significant terms for large values of x are

$$1 + \frac{\alpha_1}{Bx^p} + \dots \simeq 0,$$

so that, for closest approximation, we take

$$p = 0, \quad B = -\alpha_1.$$

Hence $\qquad y \simeq k_1 x - \alpha_1,$

eading to the asymptote

$$y - k_1 x + \alpha_1 = 0;$$

and similarly for $k_2, k_3, \dots, k_n$.

(ii) THE 'TERMS OF HIGHEST DEGREE' RULE FOR ASYMPTOTES. *If $f(x, y)$ can be expressed in the form*

$$(y - k_1 x + \alpha_1)(y - k_2 x + \alpha_2) \dots (y - k_n x + \alpha_n) + v_{n-2}(x, y) = 0,$$

where, after the product of the n linear forms $y - k_i x + \alpha_i$, the rest of the equation consists of terms of degree $n - 2$ at most, then, provided that $k_1, k_2, \dots, k_n$ are all different, the lines

$$y - k_1 x + \alpha_1 = 0,$$
$$y - k_2 x + \alpha_2 = 0,$$
$$\dots\dots\dots\dots\dots\dots$$
$$y - k_n x + \alpha_n = 0$$

are asymptotes of the curve $f(x, y) = 0$.

Consider the approximation

$$y \simeq k_1 x + B x^p \quad (p < 1).$$

On substituting, we have

$$(Bx^p + \alpha_1)\{(k_1 - k_2)x + \dots\} \dots \{(k_1 - k_n)x + \dots\} + v_{n-2}(x, k_1 x + \dots) \simeq 0,$$

or

$$(Bx^p + \alpha_1)\{(k_1 - k_2) \dots (k_1 - k_n)x^{n-1} + \dots\} + v_{n-2}(x, k_1 x + \dots) \simeq 0.$$

The terms of highest degree are (with $p < 1$)

$$(k_1 - k_2) \ldots (k_1 - k_n) Bx^{p+n-1} + (k_1 - k_2) \ldots (k_1 - k_n) \alpha_1 x^{n-1} + \ldots \eqsim 0,$$

and so (since $k_1, k_2, \ldots, k_n$ are distinct) we have for closest approximation the relations

$$p = 0, \quad B = -\alpha_1,$$

so that the approximation is

$$y \eqsim k_1 x - \alpha_1,$$

leading to the asymptote

$$y - k_1 x + \alpha_1 = 0;$$

and similarly for $k_2, k_3, \ldots, k_n$.

ILLUSTRATION 12. *To examine the asymptotes of the curve*

$$y^3 - 6xy^2 + 11x^2 y - 6y^3 + x^2 + y^2 - 2x + 3 = 0.$$

We have　　　　$u_n(x, y) \equiv (y - x)(y - 2x)(y - 3x),$

$$u_{n-1}(x, y) \equiv x^2 + y^2,$$

so that　　　　$\dfrac{u_{n-1}(x, y)}{u_n(x, y)} \equiv \dfrac{x^2 + y^2}{(y - x)(y - 2x)(y - 3x)}$

$$\equiv \frac{1}{y - x} - \frac{5}{y - 2x} + \frac{5}{y - 3x}.$$

Hence the asymptotes are

$$y - x + 1 = 0,$$
$$y - 2x - 5 = 0,$$
$$y - 3x + 5 = 0.$$

Moreover, we have the identity

$$(y - x + 1)(y - 2x - 5)(y - 3x + 5)$$
$$\equiv y^3 - 6xy^2 + 11x^2 y - 6y^3 + x^2 + y^2 + 30x - 25y - 25,$$

so that the equation of the curve may be expressed in the form

$$(y - x + 1)(y - 2x - 5)(y - 3x + 5) - 32x + 25y + 28 = 0,$$

corresponding to the 'terms of highest degree' rule just enunciated.

(iii) ASYMPTOTES PARALLEL TO THE AXES. *If the equation of a given curve is expressed in terms of descending powers of y in the form*

$$y^{n-r} w_r(x) + y^{n-r-1} w_{r+1}(x) + \ldots + w_n(x) = 0,$$

where $w_r(x), w_{r+1}(x), ..., w_n(x)$ *are polynomials in* x *of degree* $r, r+1, ..., n$, *then*, PROVIDED THAT THEY ARE ALL DISTINCT, *the* r *straight lines*

$$w_r(x) = 0$$

are asymptotes of the curve; and similarly with x, y *interchanged.*

On division by y^{n-r}, the equation assumes the form

$$w_r(x) + \frac{w_{r+1}(x)}{y} + ... + \frac{w_n(x)}{y^{n-r}} = 0,$$

which, for large values of y, approaches closely to the form

$$w_r(x) = 0.$$

ILLUSTRATION 13. *To find the asymptotes of the curve*

$$x^2 y^2 - 3xy^2 + x^2 + 2y^2 + x - y = 0.$$

Since $u_4(x, y) \equiv x^2 y^2,$

there are no asymptotes except those parallel to the axes. The coefficient of the highest power of y (namely y^2) is

$$x^2 - 3x + 2,$$

and the coefficient of the highest power of x (namely, x^2) is

$$y^2 + 1.$$

The latter expression does not give rise to real asymptotes, and so the asymptotes are

$$x - 1 = 0,$$
$$x - 2 = 0.$$

6. The radius of curvature at a multiple point. The method

given earlier (Vol. II, p. 114) for calculating curvature can be adapted to meet the case where the curve has a multiple point. For we have shown how to obtain, for each branch, an approximation to the equation of the curve in the form

$$y \simeq mx + Ax^n \quad (n > 1).$$

Hence we have, for this branch, the approximate relations

$$\frac{dy}{dx} \simeq m + nAx^{n-1},$$

$$\frac{d^2y}{dx^2} \simeq n(n-1)Ax^{n-2}.$$

At the origin, $\dfrac{d^2y}{dx^2}$ is zero if $n > 2$ and meaningless ('infinite') if $n < 2$. We may therefore confine ourselves to the normal case $n = 2$, when

$$\left.\frac{dy}{dx}\right)_0 = m, \quad \left.\frac{d^2y}{dx^2}\right)_0 = 2A.$$

The curvature κ is given by the formula

$$\kappa = \frac{\dfrac{d^2y}{dx^2}}{\left\{1 + \left(\dfrac{dy}{dx}\right)^2\right\}^{\frac{3}{2}}},$$

so that

$$\kappa = \frac{2A}{(1 + m^2)^{\frac{3}{2}}}.$$

ILLUSTRATION 14. *To find the radii of curvature of the branches at the origin of the curves*

(i) $2x^2 - 3xy + y^2 + x^2y + y^3 + x^4 = 0$,

(ii) $x^2 + x^4 - y^4 + y^5 = 0$.

(i) We obtained (p. 149) the approximations

$$y \simeq x + 2x^2,$$

$$y \simeq 2x - 10x^2.$$

For the first, $y' \simeq 1 + 4x$, $y'' \simeq 4$, so that

$$\kappa = \frac{4}{(1+1)^{\frac{3}{2}}} = \sqrt{2};$$

and, for the second, $y' \simeq 2 - 20x$, $y'' \simeq -20$, so that

$$\kappa = \frac{-20}{\{1+4\}^{\frac{3}{2}}} = -\tfrac{4}{5}\sqrt{5}.$$

(ii) We obtained (p. 152) the approximations

$$x \simeq y^2,$$

$$x \simeq -y^2.$$

Interchanging the roles of x, y we have

$$\frac{dx}{dy} = \pm 2y,$$

$$\frac{d^2x}{dy^2} = \pm 2,$$

so that

$$\kappa = \frac{\dfrac{d^2x}{dy^2}\bigg)_0}{\left\{1 + \left(\dfrac{dx}{dy}\right)^2_0\right\}^{\frac{3}{2}}}$$

$$= \pm\, 2.$$

7. The sketching of curves from their polar equations.

We now outline very briefly a few of the considerations to be kept in mind when sketching a curve of given polar equation $F(r, \theta) = 0$, where r may be positive or negative and where θ may take any positive or negative value (not necessarily restricted to an interval of 2π). We confine our attention to examples in which the equation may be solved to give r as a single-valued function of θ in the form

$$r = f(\theta).$$

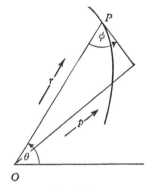

Fig. 145.

It seems harder to give a systematic treatment for polar coordinates than for Cartesians, and there is much scope for ingenuity. A routine first step may well be to construct a table of values of r for particular values of θ; for example, if $f(\theta)$ involves the trigonometric functions, it is useful to put $\theta = 0, \frac{1}{6}\pi, \frac{1}{4}\pi, \frac{1}{3}\pi, \frac{1}{2}\pi$ and so on.

The formulae which we obtained in Volume II, Chapter X, may be used, and one or two of the results are repeated here for convenience.

Let P be the point (r, θ) of the curve. The direction of the tangent at P is determined by the angle ϕ 'behind' the radius OP; if a radius vector centred on P is imagined to rotate about P from the position PO until it is first in line with the tangent at P, the angle of rotation is denoted by ϕ, and it is proved that

$$\tan\phi = r\frac{d\theta}{dr}.$$

This relation defines ϕ uniquely, since it lies between the values $0, \pi$.

For the curvature κ, we obtained the formula

$$\kappa = \left\{ r^2 + 2\left(\frac{dr}{d\theta}\right)^2 - r\frac{d^2r}{d\theta^2}\right\} \bigg/ \left\{ r^2 + \left(\frac{dr}{d\theta}\right)^2\right\}^{\frac{3}{2}},$$

the square root implied in the denominator being positive. Now (Vol. II, p. 114) the sign of κ is the same as the sign of $\dfrac{d^2y}{dx^2}$ *when x is the parameter defining the curve*; and (Vol. I, p. 54) the concavity is 'upwards' or 'downwards' according as $\dfrac{d^2y}{dx^2}$ is positive or negative.

In the present work, the parameter is θ, and the relationship connecting the concavity with the sign of the above formula for κ is more complex. It remains true, however, that the curve has an *inflexion* where $\kappa = 0$, that is, where $r^2 + 2\left(\dfrac{dr}{d\theta}\right)^2 - r\dfrac{d^2r}{d\theta^2} = 0$.

Note that the expression on the left assumes a simpler form on replacing r by its reciprocal $u \equiv 1/r$. Then

$$\frac{dr}{d\theta} = -\frac{1}{u^2}\frac{du}{d\theta}, \quad \frac{d^2r}{d\theta^2} = -\frac{1}{u^2}\frac{d^2u}{d\theta^2} + \frac{2}{u^3}\left(\frac{du}{d\theta}\right)^2,$$

so that the expression is, on substituting and simplifying,

$$\left(\frac{d^2u}{d\theta^2} + u\right)\bigg/ u^3. \tag{1}$$

In particular, the points of inflexion satisfy the equation

$$\frac{d^2u}{d\theta^2} + u = 0,$$

provided that the expression (1) changes sign for such a value of θ.

Singularities

The location of singularities is, of course, very helpful. The origin will certainly be a multiple point if the equation $f(\theta) = 0$ has more than one (significantly) distinct solution. Suppose, more generally, that there is a singularity at the point (ρ, α), where $\rho = f(\alpha)$. Since r is a single-valued function of θ, this can happen only if EITHER the value of r corresponding to a value $\alpha \pm 2\pi$, $\alpha \pm 4\pi$, $\alpha \pm 6\pi$, ... of θ is also ρ, OR if the value of r corresponding to a value $\alpha \pm \pi$, $\alpha \pm 3\pi$, $\alpha \pm 5\pi$, ... of θ is $-\rho$. For example, if $f(\theta)$ is a trigonometrical function of θ, it is often (but not always) sufficient to consider the points, if any, expressed in the alternative forms (ρ, α), $(-\rho, \alpha + \pi)$; and these are found by solving the equation

$$f(\alpha) + f(\alpha + \pi) = 0.$$

Thus, for the curve $\qquad r = \sin\theta + \cos 2\theta$,

we should consider the equation

$$\{\sin\theta + \cos 2\theta\} + \{\sin(\theta + \pi) + \cos(2\theta + 2\pi)\} = 0$$

or $\qquad\qquad\qquad \cos 2\theta = 0$,

giving $\qquad\qquad \theta = \dfrac{\pi}{4},\ \dfrac{3\pi}{4},\ \dfrac{5\pi}{4},\ \dfrac{7\pi}{4},\ \ldots$

and singularities at the two points

$$\left(\frac{1}{\sqrt{2}}, \frac{\pi}{4}\right) \equiv \left(-\frac{1}{\sqrt{2}}, \frac{5\pi}{4}\right) \quad \text{and} \quad \left(\frac{1}{\sqrt{2}}, \frac{3\pi}{4}\right) \equiv \left(-\frac{1}{\sqrt{2}}, \frac{7\pi}{4}\right).$$

Asymptotes

The identification of the (linear) asymptotes is a fairly simple process, but the justification of the method seems less easy. For convenience, we write $1/r = u$, $1/f(\theta) = g(\theta)$, so that the equation of the curve assumes the form

$$u = g(\theta).$$

For large values of r, its reciprocal u is small, and so we begin by finding the solutions, if any, of the equation $g(\theta) = 0$. Let $\theta = \alpha$ be such a solution, where

$$g(\alpha) = 0.$$

Take an associated system of Cartesian coordinates in which the positive x, y axes are the radii $\theta = 0$, $\tfrac{1}{2}\pi$ respectively. Let β be a value of θ near to α, and let $Q(b, \beta)$ be the corresponding point of the curve (fig. 146), so that

$$\frac{1}{b} = g(\beta).$$

The line through Q in the direction α is given by the equation

$$x \sin\alpha - y\cos\alpha = b\sin(\alpha - \beta)$$

$$= \frac{\sin(\alpha - \beta)}{g(\beta)}.$$

Now suppose that β approaches the value α, and suppose that, as it does so,

$$\lim_{\beta \to \alpha} \frac{\sin(\alpha - \beta)}{g(\beta)} = h.$$

Then the line takes up the limiting position given by the equation

$$x\sin\alpha - y\cos\alpha = h.$$

The distance of the point Q $(b\cos\beta, b\sin\beta)$ from this line, being

$$\pm\,(b\cos\beta\sin\alpha - b\sin\beta\cos\alpha - h)$$
$$=\pm\{b\sin(\alpha-\beta)-h\}$$
$$=\pm\left\{\frac{\sin(\alpha-\beta)}{g(\beta)}-h\right\},$$

tends to zero as β tends to α. Thus the line whose equation is

$$x\sin\alpha - y\cos\alpha = \lim_{\beta\to\alpha}\frac{\sin(\alpha-\beta)}{g(\beta)}$$

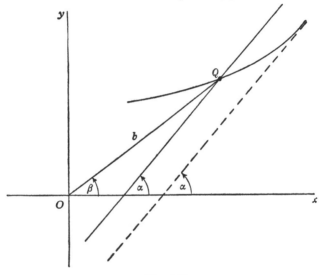

Fig. 146.

is an *asymptote* of the curve, being approached ever more closely by the point Q of the curve as β tends to α.

In accordance with the usual rule for the limiting value of the ratio of two functions each of which tends to zero, we have the relation

$$\lim_{\beta\to\alpha}\frac{\sin(\alpha-\beta)}{g(\beta)} = -\left.\frac{\cos(\alpha-\beta)}{g'(\beta)}\right|_{\beta=\alpha}$$
$$=\frac{-1}{g'(\alpha)}.$$

Thus, in normal cases, the equation of the asymptote is

$$x\sin\alpha - y\cos\alpha = \frac{-1}{g'(\alpha)}.$$

ILLUSTRATION 15. *To find the asymptotes of the curve*

$$r(\cos\theta - \sin\theta) = 1 + \cos\theta + \sin\theta.$$

If $1/r = u$, then
$$u = \frac{\cos\theta - \sin\theta}{1 + \cos\theta + \sin\theta}.$$

The asymptotes are obtained from the equation
$$\cos\theta - \sin\theta = 0,$$

so that $\theta = \tfrac{1}{4}\pi, \tfrac{5}{4}\pi, \ldots$

If $\theta = \tfrac{1}{4}\pi$, then, since
$$g'(\theta) = \frac{-\sin\theta - \cos\theta}{(1 + \cos\theta + \sin\theta)} - \frac{(\cos\theta - \sin\theta)^2}{(1 + \cos\theta + \sin\theta)^2},$$

we have
$$g'(\tfrac{1}{4}\pi) = -\sqrt{2}/(1 + \sqrt{2}),$$

and so the equation of the asymptote is
$$\frac{x}{\sqrt{2}} - \frac{y}{\sqrt{2}} = \frac{1 + \sqrt{2}}{\sqrt{2}},$$

or
$$x - y = \sqrt{2} + 1.$$

The value $\theta = \tfrac{5}{4}\pi$ gives $x - y = -\sqrt{2} + 1$, and other values of θ give repetitions of these two asymptotes.

ILLUSTRATION 16. *To find the asymptotes of the curve*

$$r\sqrt{\{\cos\theta(\cos\theta - \sin\theta)\}} = 1.$$

If $1/r = u$, then $u = \sqrt{\{\cos\theta(\cos\theta - \sin\theta)\}}.$

The asymptotes are obtained from the equations
$$\cos\theta = 0, \quad \cos\theta - \sin\theta = 0,$$

so that $\theta = \tfrac{1}{2}\pi, \tfrac{3}{2}\pi, \ldots; \quad \theta = \tfrac{1}{4}\pi, \tfrac{5}{4}\pi, \ldots$

If $\theta = \tfrac{1}{2}\pi$, then
$$\lim_{\beta \to \frac{1}{2}\pi} \frac{\sin(\tfrac{1}{2}\pi - \beta)}{\sqrt{\{\cos\beta(\cos\beta - \sin\beta)\}}} = \lim_{\beta \to \frac{1}{2}\pi} \sqrt{\left\{\frac{\cos\beta}{\cos\beta - \sin\beta}\right\}} = 0.$$

Similarly for $\theta = \tfrac{3}{2}\pi$. Thus one asymptote is the line

$$x = 0.$$

If $\theta = \tfrac{1}{4}\pi$, then
$$\lim_{\beta \to \frac{1}{4}\pi} \frac{\sin(\tfrac{1}{4}\pi - \beta)}{\sqrt{\{\cos\beta(\cos\beta - \sin\beta)\}}} = \lim_{\beta \to \frac{1}{4}\pi} \frac{\frac{1}{\sqrt{2}}(\cos\beta - \sin\beta)}{\sqrt{\{\cos\beta(\cos\beta - \sin\beta)\}}}$$
$$= \lim_{\beta \to \frac{1}{4}\pi} \sqrt{\left\{\frac{\cos\beta - \sin\beta}{2\cos\beta}\right\}} = 0.$$

Similarly for $\theta = \frac{5}{4}\pi$. Thus the other asymptote is the line

$$x - y = 0.$$

(We could not use the formula involving $g'(\alpha)$, since $g'(\frac{1}{2}\pi)$, $g'(\frac{1}{4}\pi)$ are both 'infinite'. The answer would have been correct, but the justification might not.)

Note. It may be remarked that, in general, if the equation of the curve occurs in the form

$$rG(\theta) = F(\theta),$$

then $$g(\theta) = \frac{G(\theta)}{F(\theta)},$$

$$g'(\theta) = \frac{G'(\theta)}{F(\theta)} - \frac{G(\theta)\,F'(\theta)}{\{F(\theta)\}^2}.$$

Hence the asymptotes occur when $\theta = \alpha$, where

$$G(\alpha) = 0,$$

so that $$g'(\alpha) = \frac{G'(\alpha)}{F(\alpha)},$$

and the equation for the asymptote is

$$x \sin \alpha - y \cos \alpha = \frac{-F(\alpha)}{G'(\alpha)}.$$

This is the form often given, and it can be useful; but it needs care. Thus the curve

$$r = \frac{1}{\sqrt{\{\cos\theta(\cos\theta - \sin\theta)\}}}$$

just discussed leads us, with

$$F(\theta) = 1, \quad G(\theta) = \sqrt{\{\cos\theta(\cos\theta - \sin\theta)\}},$$

into theoretical difficulties. On the other hand, with

$$F(\theta) = \sqrt{\{\cos\theta(\cos\theta - \sin\theta)\}}, \quad G(\theta) = \cos\theta(\cos\theta - \sin\theta),$$

we avoid these difficulties, at the expense, perhaps, of some mental confusion about the apparently arbitrary selections possible for $F(\theta)$ and $G(\theta)$.

Finally, it should be remembered that the curve

$$r = f(\theta)$$

may have a *circular asymptote* if $f(\theta)$ tends to a finite limit as θ tends to infinity. Thus, if

$$\lim_{\theta \to \infty} f(\theta) = a,$$

then the curve (of 'spiral' shape) approaches asymptotically the circle
$$r = a.$$

For example if $\qquad r = \tanh\theta,$

then, since $\qquad \lim_{\theta \to \infty} \tanh\theta = \lim_{\theta \to \infty} \dfrac{e^\theta - e^{-\theta}}{e^\theta + e^{-\theta}} = 1,$

the asymptotic circle is $\qquad r = 1.$

REVISION EXAMPLES XIV

'Scholarship' Level

1. The equation of a curve is
$$x^2 y^2 - x^2 + y^2 = 0.$$
(i) Find the equations of the tangents at the origin.
(ii) Find the equations of the real asymptotes.
(iii) Show that the numerical value of y is always less than that of x.
(iv) Show that the numerical value of y is always less than unity
(v) Sketch the curve.

2. The equation of a curve is
$$(x^2 + y^2)^2 = x^2(1 + y^2).$$
(i) Discuss the symmetry of the curve with respect to the origin and the axes.
(ii) Determine the coordinates of the points at which the tangents are parallel to the x-axis.
(iii) Prove that the curve has no real asymptotes.
(iv) Find the points of intersection of the curve with the axes, and show that x is always numerically not greater than 1.
(v) Sketch the curve.

3. Find the asymptotes of the curve
$$x^3 - xy^2 - y = 0,$$
and prove that the origin is a point of inflexion.

4. Find the tangents at the origin and the asymptotes to the curve
$$x(x^2 - y^2) = y(x + 2y).$$
Prove that each asymptote meets the curve at a finite point which lies on the straight line
$$13x + 8y = 6.$$

5. Trace the curve
$$x^3 + xy^2 + x^2 - y^2 = 0.$$
Draw the tangents to the curve at the origin and find the equation of the real asymptote.

6. Find the asymptote of the curve
$$x^3 - y^3 = ay(x + y).$$

There are two other tangents to this curve which are parallel to the asymptote. Find their equations and the coordinates of their points of contact.

Trace the curve.

7. Sketch roughly the curve
$$6y^2 = x(x - y)(x - 3y);$$
find its asymptotes and indicate them on your sketch.

Show that each asymptote meets the curve in a single finite point and that the three points so obtained are collinear.

8. Find the asymptotes of the curve
$$(x + y - 1)^3 = x^3 + y^3,$$
and prove that they meet the curve only at infinity.

Prove also that there is no point on the curve for values of x between λ and 1, where λ is the real root of the equation
$$3\lambda^3 + 3\lambda^2 - 3\lambda + 1 = 0.$$
Give a sketch showing the general form of the curve.

9. Find the equation of the straight line which is asymptotic to the curve
$$x^2(x - y) + y^2 = 0.$$

Prove also the following facts and give a sketch of the curve:
  (i) the origin is a cusp;
  (ii) no part of the curve lies between $x = 0$ and $x = 4$;
  (iii) the curve consists of two infinite branches, one lying in the first quadrant and the other in the second and third quadrants.

10. Trace the curve
$$(x^2 - y^2)^2 - 4y^2 + y = 0.$$

11. Find the asymptotes of the curve
$$(y - x)^2(y + x) - (y - x)(4y - x) + 2y - x = 0.$$
Investigate on which side of each asymptote the curve lies, and trace the curve.

12. Find the asymptotes of the curve
$$(x+y)(x-2y)(x-y)^2 + 3xy(x-y) + x^2 + y^2 = 0.$$
Determine on which side the curve approaches each of the ends of the parallel asymptotes.

13. Find the asymptotes of the curve
$$x^3 + 2x^2 + x = xy^2 - 2xy + y.$$
Show that there are values which $y-x$ never takes, and sketch the curve.

14. Find the asymptotes of the curve
$$2x^3 - 5x^2y - xy^2 + 6y^3 - 4x^2 - 4xy + 15y^2 - 6x + 9y + 1 = 0.$$
Give a rough sketch of the curve, indicating clearly on which side of the asymptotes the curve lies.

15. Give a rough sketch of the curve
$$x^3 - xy^2 - 2x^2 + 3xy + y^2 - 3y = 0,$$
and indicate clearly on which sides of the asymptotes the various branches of the curve lie.

16. Trace the curve
$$2xy^2 + 2(x^2 - x + 2)y - (x^2 - 5x + 2) = 0.$$
Prove that at no finite real point of the curve is the tangent parallel to the x-axis.

17. Determine the asymptotes of the curve
$$(y-1)^2(y^2 - 4x^2) = 3xy.$$
Investigate on which sides of the asymptotes corresponding branches of the curve lie, and trace the curve.

18. Trace the curve
$$y^2(x^2 + 2y^2) = 2(x^2 - 4x + 2y^2)^2.$$

19. Show that $x-y = 3$ is an asymptote of
$$(x-y+1)(x-y-2)(x+y) = 8x - 1,$$
find the other asymptotes, and sketch the curve.

20. Find the asymptotes of the curve
$$x^2(x+y) = x + 4y,$$
and trace the curve.

21. Trace the curve
$$x^5 - a^2x^2y - b^2xy^2 + y^5 = 0.$$
What modification occurs when $a^2 = b^2$?

22. Trace the curve
$$(x^2 - y^2)^2 - a^3x = 0.$$

23. Find the asymptotes of the curve
$$x^4 + 3x^3y + 2x^2y^2 + 2xy + 3x + y = 0;$$
determine on which side or sides the curve approaches each asymptote, and where it cuts the asymptotes.

24. Trace the curve
$$y^3 - 3x^2y + 4a(x^2 - y^2) = 0,$$
finding its asymptotes and the curvatures at the points where $x = 0$ meets it.

25. Find the equation of the tangent at the point $(1, 2)$ to the curve given by
$$xy(x + y) = x^2 + y^2 + 1,$$
and determine the point at which it intersects the curve. Find the asymptotes and trace the curve.

26. The lines whose equations are $x = y$, $x = -y$, $x = 2y$ are the asymptotes of a cubic curve which touches the axis of x at the origin and which passes through the point $(0, b)$. What is the equation of the curve?

27. Sketch the curve $x^2 - y^2 = y^3$.
Find (i) the position of the centre of curvature of either branch of the curve at the origin, and (ii) the area of the loop.

28. Sketch the curve $xy = x^3 + y^3$,
and find (i) the radii of curvature at the origin of coordinates, (ii) the area of the loop.

29. Investigate the curvature at the point $(1, 2)$ of the curve
$$(y - 2)^2 = x(x - 1)^2.$$

30. Trace the curve
$$4(x^2 + 2y^2 - 2ay)^2 = x^2(x^2 + 2y^2),$$
and find the radii of curvature of the two branches at the origin.

31. Find the asymptotes of the curve
$$x^2y + xy^2 = x^2 - 4y^2,$$
and trace it.

Find the cubic which has $x + y = 1$ as an asymptote and touches both axes at the origin, the radii of curvature there being 1 and 2 units in length.

32. Find the asymptotes of the curve
$$y^2 = \frac{a^3x}{a^2 - x^2}$$
and find the radius of curvature at the origin.

Sketch the curve.

33. Find the curvature at the origin of each of the branches of the curve
$$x^3 - x^2y - xy^2 + y^3 = 4xy - 2y^2,$$
and trace the curve.

34. Find the radii of curvature at the origin of the curve
$$2x^2 + xy - 3y^2 + 2x^3 - x^2y + 4y^4 = 0.$$

35. Evaluate $\displaystyle\int_a^b \sqrt{\{(b-x)/(x-a)\}}\,dx \quad (a < b)$

by means of the substitution $x = a\sin^2\theta + b\cos^2\theta$, or otherwise.

Make a rough drawing of the curve
$$x^3 + 3xy^2 - 3a(x^2 - y^2) = 0,$$
and show that the area of its loop is $3a^2$.

36. Find the coordinates of the node of the curve
$$(x + y + 1)y + (x + y + 1)^3 + y^3 = 0,$$
and the area of the loop at the node.

37. Make a rough sketch of the curve
$$x(x^2 + y^2 - 3a^2) = 2a^3,$$
and prove that the area between the curve and its asymptote is $3\pi a^2$.

38. Sketch roughly the curve
$$y^2(a^2 + x^2) = x^2(a^2 - x^2),$$
and find the area of one of its loops.

M III

39. Trace the curve
$$(x^2+y^2)^2 = 16axy^2,$$
and find the areas of its loops.

Prove that the smallest circle that will completely circumscribe the curve has radius $3\sqrt{3}\,a$.

40. Trace the curve $y^2 = x^2(x-a)$

for $a = 1, 0, -1$. Find the area enclosed by the loop in the case $a = -1$.

41. Trace the curve
$$16a^3y^2 = b^2x^2(a-2x),$$
where a, b are positive, and find the area enclosed by the loop.

If $16a^2 = 3b^2$, show that the perimeter of the loop is $\frac{1}{2}b$.

42. Trace the curve $y^2 = \dfrac{x^2(3-x)}{1+x}$,

and find the area of the loop.

43. Trace the curve
$$x^3+y^3-3axy = 0,$$
and find the area of the loop.

44. Give a rough sketch of the curve
$$x^5+y^5 = 5ax^2y^2.$$
Find the area of the loop of the curve.

45. Sketch the curve whose equation in polar coordinates is
$$r = 1+\cos 2\theta.$$
Prove that the length of the curve corresponding to $0 \leqslant \theta \leqslant 2\pi$ is
$$8+\frac{4}{\sqrt{3}}\log(2+\sqrt{3}).$$

46. Sketch the curve whose polar equation is $r^2 = a^2(1+3\cos\theta)$ and find the area it encloses.

47. Trace the curve $r = a(2\cos\theta - 1)$, find the areas of its loops, and show that their sum is $3\pi a^2$.

48. Trace the curve $r = a(\cos\theta + \cos 2\theta)$, and show that the curve crosses itself at the points where $r = \frac{1}{2}a\sqrt{2}$, $\theta = \pm\frac{1}{4}\pi$.

Prove that the area of that portion of the largest loop that is not common to the other loops is $\frac{4}{3}a^2\sqrt{2}$.

49. Trace the curve $r = 2 + 3\cos 2\theta$, and find the areas of the loops.

50. Trace the curve $r = a(\sin\theta - \cos 2\theta)$, and find the area of the loop which passes through the point $(2a, \tfrac{1}{2}\pi)$.

51. Sketch the curve

$$r(1 - 2\cos\theta) = 3a\cos 2\theta,$$

and find the equations of its asymptotes.

52. Trace the curve

$$r\cos\theta + a\cos 2\theta = 0.$$

Show that the area of the loop is $a^2(2 - \tfrac{1}{2}\pi)$, and that the area enclosed between the curve and its asymptote is $a^2(2 + \tfrac{1}{2}\pi)$.

53. Determine the asymptotes of the curve $r\cos 3\theta = a$, and sketch the curve.

54. Sketch the curve $r(\cos\theta + \sin\theta) = a\sin 2\theta$, and find the area of the loop of the curve.

55. Trace the curve $r\cos\theta = a\sin 3\theta$, and prove that the area of a loop is $\tfrac{1}{8}a^2(9\sqrt{3} - 4\pi)$.

ANSWERS TO EXAMPLES

CHAPTER XIV

Examples I:

    1. Continuous.   **2, 3, 4.** Discontinuous.

Examples II:

    1. $2xy^5,\ 5x^2y^4$.   **2.** $3x^2y^{-2},\ -2x^3y^{-3}$.

    3. $\sin y,\ x\cos y$.   **4.** $-y\sin xy,\ -x\sin xy$.

    5. $e^x\cos 2y,\ -2e^x\sin 2y$.   **6.** $e^x,\ e^y$.

    7. $1,\ 0$.

    8. $3(1+x)^2\,e^{-xy}-y(1+x)^3\,e^{-xy},\ -x(1+x)^3\,e^{-xy}$.

    9. $\tfrac{1}{2}y/\sqrt{x},\ \sqrt{x}$.

    10. $3x^2y^4z^{-2},\ 4x^3y^3z^{-2},\ -2x^3y^4z^{-3}$.   **11.** $4x^3,\ 3y^2,\ 2z$.

    12. $e^x\sin y\cos z,\ e^x\cos y\cos z,\ -e^x\sin y\sin z$.

    13. $\cos x\sin y\sin z,\ \sin x\cos y\sin z,\ \sin x\sin y\cos z$.

    14. $3(1+x)^2\,e^{yz},\ z(1+x)^3\,e^{yz},\ y(1+x)^3\,e^{yz}$.

    15. $\tan^{-1}yz,\ \dfrac{xz}{1+y^2z^2},\ \dfrac{xy}{1+y^2z^2}$.

Examples IV:

    1. $2y,\ 2x,\ 2r\sin 2\theta,\ 2r\cos 2\theta$.

    2. $3x^2,\ -3y^2,\ 3(x^3-y^3)/r,\ -3xy(x+y)/r$.

Examples VI:

    1. $16t^{15}$.   **2.** $(\cos^2 t-\sin t)\,e^{\sin t}$.

    3. 0.   **4.** $2t^3(2+3t^3)\,e^{2t^3}$.

Examples VII:

    1. $\cos 2\theta\sin^2\phi\cos\phi,\ \sin\theta\cos\theta\sin\phi(2\cos^2\phi-\sin^2\phi)$.

    2. $2(x+y+z)-2xy^2z^2$, etc.

    3. $e^{x+y}\{\cos(x-y)-\sin(x-y)\},\ e^{x+y}\{\cos(x-y)+\sin(x-y)\}$.

    4. $6u(u^2+v^2)^2\log(u^2-v^2)+\dfrac{2u(u^2+v^2)^3}{u^2-v^2}$,

$$6v(u^2+v^2)^2\log(u^2-v^2)-\dfrac{2v(u^2+v^2)^3}{u^2-v^2}.$$

Examples IX:

1. 6.4%.   2. 17.6%.   3. 11.1%.

Examples X:

1. $2y^2$, $4xy$, $2x^2$, $4y$, $4x$.
2. $6xy^4$, $12x^2y^3$, $12x^3y^2$, $24xy^3$, $36x^2y^2$.
3. $e^x \sin y$, $e^x \cos y$, $-e^x \sin y$, $e^x \cos y$, $-e^x \sin y$.
4. 0, y^{-1}, $-xy^{-2}$, 0, $-y^{-2}$.
5. $2\sin y$, $2x\cos y$, $-x^2 \sin y$, $2\cos y$, $-2x\sin y$.
6. 0, $-e^{-v}$, $x\,e^{-v}$, 0, e^{-v}.   7. 0, $-y^{-2}$, $2xy^{-3}$, 0, $2y^{-3}$.
8. $-\dfrac{2xy}{(1+x^2)^2}$, $\dfrac{1}{(1+x^2)}$, 0, $-\dfrac{2x}{(1+x^2)^2}$, 0.
9. $(x^2+4x+2)\,y^2\,e^x$, $(2x^2+4x)\,y\,e^x$, $2x^2\,e^x$,
$$(2x^2+8x+4)\,y\,e^x,\ (2x^2+4x)\,e^x.$$

Examples XI:

1. $\dfrac{2a}{y}$, $-\dfrac{4a^2}{y^3}$.   2. $-\dfrac{\alpha x}{\beta y}$, $-\dfrac{\alpha}{\beta^2 y^3}$.
3. $-\dfrac{x^2}{y^2}$, $-\dfrac{2a^3 x}{y^5}$.
4. $-\dfrac{2xy+y^2}{2xy+x^2}$, $\dfrac{6y(x+y)(x^2+xy+y^2)}{x^2(x+2y)^3}$.

REVISION EXAMPLES X

8. $v(x,y) \equiv e^x(x\sin y + y\cos y)$, $F(z) \equiv z\,e^z$.
26. $-\dfrac{f_u g_v - f_v g_u}{f_x g_v - f_y g_x}$.
36. $\frac{1}{2}b\sin C$, $b\sin C$, $\frac{1}{2}a\cot A$.
44. (i) 2,   (ii) 1.
45. $u \equiv ax+b$, $v \equiv acy+d$, $w \equiv cz$.   46. $A\sinh v$.
49. $b\left\{\dfrac{h}{a} + (\cot B + 2\cot A)\,\alpha\right\}$.
50. $p(y) \equiv (ay+b)(c'y+d') - (a'y+b')(cy+d)$.
     Take $y \equiv e^{\int p^{-1}\,dy}$, with similar form for X.

REVISION EXAMPLES XI

1. Max. $\frac{1}{4}$ at $(2,1)$; min. $-\frac{3}{4}$ at $(-\frac{2}{3}, -\frac{1}{3})$.
3. 8, 4, 8.   4. $c^2/(a^2+b^2+1)$.
5. Max. at (λ, λ) if $\lambda^2 > a^2$; min. at (λ, λ) if $\lambda^2 < a^2$; no stationary value at $\lambda^2 = a^2$.
6. $\frac{1}{6}\sqrt{3}\,abc$.
7. Roots of $\Sigma l^2(br^2-1)(cr^2-1) = 0$.
8. Min $\frac{5}{9}a^3$ at points such as $(-\frac{1}{3}a, \frac{2}{3}a, \frac{2}{3}a)$.
      $[(a,0,0),\ (0,a,0),\ (0,0,a)$ are not genuine points.]
11. $\frac{1}{14}k$, $\frac{1}{7}k$.
12. Greatest is $(\Sigma a^{2p/(p-2)})^{(p-2)/p}$. Least is the least of a, b, c.
14. 1, $\frac{1}{2}$, $-\frac{1}{2}$.
16. Max. 27 at $(2, -1, -4)$; min. 3 at $(0, -1, -2)$.
19. Neither max. nor min. at $(-5, 4, 4)$; min. at $(1, 1, 1)$.
20. $\dfrac{abc}{bc+ca+ab}$, $1+\dfrac{(p-1)(q-1)(r-1)}{\Sigma(q-1)(r-1)}$.
23. $\dfrac{x}{m} = \dfrac{y}{n} = \dfrac{z}{p} = \dfrac{A}{m+n+p}$.

CHAPTER XVI

Examples II:

1. r.   2. e^{2u}.   3. r.   4. 0.   5. $e^{3t}\sin\theta$.

Examples III:

2. No.   3. Yes.   4. $a = -3$.

REVISION EXAMPLES XII

1. $\dfrac{\lambda-\mu}{4\sqrt{\{-(a+\lambda)(a+\mu)(b+\lambda)(b+\mu)\}}}$.
7. $\dfrac{2(x+2y+1)(z+1)}{x+y+2}$.
8. $2v\,e^{u^2+v^2}/(1+3uv)$.

CHAPTER XVII

Examples I:

1. $\frac{1}{6}\rho$.   2. $\frac{5}{16}\pi k$.   4. $\frac{32}{15}\pi k$.

REVISION EXAMPLES XIII

1. $\pi - 4$.

2. $(\frac{8}{15}a, \frac{8}{15}a)$.

3. $0; (e^{ab} - 1 - ab)/b$.

4. $\frac{27}{8}a^4, \frac{8}{3}a^4, \frac{65}{24}a^4$.

6. $\frac{1}{24}a^4b^2$.

7. $\frac{1}{24}$.

8. $\log 2 + \frac{89}{480}$.

10. πa^3.

11. $\frac{1}{2}\pi abc$.

12. $\frac{2048}{75}a^{\frac{5}{2}}$.

13. $u = y^2/x, v = x^2/y$.

14. $2(3\pi - 4)a^3/9$.

15. $\frac{1}{14}\pi a^5 b^5$.

18. $\frac{4}{15}\pi(a+b+c)$.

20. $\int_{\frac{1}{4}\pi}^{\frac{1}{2}\pi} d\theta \left\{ \int_0^{4a \operatorname{cosec}\theta \cot\theta} f(r\cos\theta, r\sin\theta)\, r\, dr \right\}; \frac{4}{3}(3\pi - 10)a^2$.

21. $\frac{4}{15}\pi abc(b^2 + c^2)$.

25. $\frac{1}{12}a^3 b\rho \left(\pi + \dfrac{45\sqrt{3}}{56} \right)$.

CHAPTER XVIII

Examples III:

1. $y = \pm x(1 - \frac{1}{2}x)$.

2. $y = \pm x(1 - 2x^2)$.

3. $x = \pm y(1 + 2y)$.

4. $x = \pm y(1 - 2y^2)$.

5. $y = 2x - \frac{1}{2}x^2$ or $y = \frac{1}{2}x^2$.

6. $y = -4x - x^3$ or $y = x^3$.

Examples IV:

1. $2x - y + 1 = 0, x - y + 1 = 0$.

2. $x - y = 0, x + y + 2 = 0$.

3. $x + y + \frac{4}{3} = 0, x - y + 2 = 0, 2x - y - \frac{8}{3} = 0$.

4. $x + y + \frac{1}{2} = 0, x - y + \frac{3}{2} = 0, x - y - 1 = 0$.

5. $x - y + 2 = 0, x + y = 0, x + y + 2 = 0$.

Examples V:

1. $2(y + x - \frac{1}{4})^2 + 3x = 0$.

2. $(2x + \frac{1}{4})^2 = y$.

3. $(y + 3x + 3)^2 = 9x$.

REVISION EXAMPLES XIV

1. (i) $x \pm y = 0$,   (ii) $y \pm 1 = 0$.

2. (ii) $(0, 0), \left(\pm \dfrac{1}{\sqrt{3}}, \pm \dfrac{1}{\sqrt{3}} \right)$,   (iv) $(0, 0), (\pm 1, 0)$.

3. $x = 0$, $x \pm y = 0$.

4. Tangents $y = 0$, $x + 2y = 0$;

   Asymptotes $x + 2 = 0$, $x + y - \frac{1}{2} = 0$, $x - y - \frac{3}{2} = 0$.

5. $x = 1$.

6. $x - y = \frac{2}{3}a$; $x - y - a = 0$ at $(0, -a)$ and $x - y + \frac{1}{3}a = 0$

   at $\left(-\frac{2}{9}a, \frac{1}{9}a\right)$.

7. $x = 2$, $x - y + 3 = 0$, $x - 3y - 1 = 0$.

8. $x = 1$, $y = 1$, $x + y = 0$.   9. $x - y + 1 = 0$.

10. Asymptotes are $x \pm y \pm 1 = 0$.

11. $x - y + 1 = 0$, $x - y + \frac{1}{2} = 0$, $x + y - \frac{5}{2} = 0$.

12. $x + y - \frac{1}{2} = 0$, $x - 2y + 2 = 0$, $x - y - 2 = 0$, $x - y + \frac{1}{2} = 0$.

13. $x = 0$, $x - y + 2 = 0$, $x + y = 0$;

   $y - x$ is not between $-2, 0$ or $2, 4$.

14. $x + y + 1 = 0$, $x - 2y - 3 = 0$, $2x + 3y = 0$.

15. Asymptotes are $x - 1 = 0$, $x + y - 2 = 0$, $x - y + 1 = 0$.

16. Asymptotes are $x = 0$, $y = \frac{1}{2}$, $x + y - \frac{1}{2} = 0$.

17. $y - 1 = 0$, $y \pm 2x = 0$.   19. $x - y + 2 = 0$, $x + y = 0$.

20. $x + y = 0$, $x \pm 2 = 0$.

21. Asymptote is $x + y = 0$. If $a^2 = b^2$, the line $x + y = 0$ is part of the locus.

22. Asymptotes $x \pm y = 0$.

23. $x = 0$, cutting at $(0, 0)$; $x + y = 0$, cutting at $(0, 0)$, $(1, -1)$,

   $x + 2y = 0$, cutting at $(0, 0)$, $\left(\frac{5}{2}, -\frac{5}{4}\right)$.

24. Asymptotes $y = \frac{4}{3}a$, $\pm 3\sqrt{3}\,x - 3y + 4a = 0$; curvature

   $-\frac{1}{8}\sqrt{2}\,a^{-1}$ each branch at origin and a^{-1} at $(0, 4a)$.

25. $6x + y = 8$, $\left(\frac{13}{6}, -5\right)$; asymptotes $x = 1$, $y = 1$, $x + y + 2 = 0$.

26. $(x^2 - y^2)(x - 2y) = 2b^2y$.

27. $(2, -2)$, $(-2, -2)$; $\frac{8}{15}$.

28. $\frac{1}{2}$; $\frac{1}{6}$.   29. $\frac{1}{4}\sqrt{2}$, $-\frac{1}{4}\sqrt{2}$.   30. $2a$, $\frac{2}{3}a$.

31. $x + 4 = 0$, $y - 1 = 0$, $x + y - 3 = 0$;

   $4xy = 2x^3 + 9x^2y + 8xy^2 + y^3$.

32. $x = \pm a$, $y = 0$; $\frac{1}{2}a$.

33. $\frac{1}{2}$ for $y = 0$; $-\frac{3}{50}\sqrt{5}$ for $y = 2x$. The asymptote is

   $x + y = -\frac{3}{2}$, and asymptotic parabola $\left(y - x + \frac{1}{4}\right)^2 = x$.

34. $5\sqrt{2}$, $\frac{65}{144}\sqrt{13}$.   35. $\frac{1}{2}\pi(b - a)$.

36. $(-1, 0)$; $\frac{1}{6}$.   37. The asymptote is $x = 0$.

38. $(\frac{1}{2}\pi - 1)a^2$.   39. $4\pi a^2$.

40. $\frac{8}{15}$.   41. $\frac{1}{30}ab$.

42. $3\sqrt{3}$.   43. $\frac{3}{2}a^2$.

44. $\frac{5}{2}a^2$.   46. $a^2\{\pi + 2\sqrt{2} - \sin^{-1}(\frac{2}{3}\sqrt{2})\}$.

47. $a^2(\pi - \frac{3}{2}\sqrt{3})$, $a^2(2\pi + \frac{3}{2}\sqrt{3})$.

49. $\frac{17}{2}\tan^{-1}(\sqrt{5}) + \frac{3}{2}\sqrt{5}$; $\frac{17}{4}\pi - \frac{17}{2}\tan^{-1}(\sqrt{5}) - \frac{3}{2}\sqrt{5}$.

50. $(\frac{1}{3}\pi + \frac{9}{16}\sqrt{3})a^2$.   51. $x\sqrt{3} \pm y = a\sqrt{3}$.

53. $3x + a = 0$, $x \pm y\sqrt{3} = \frac{2}{3}a$.   54. $\frac{1}{4}a^2(\pi - 2)$.

INDEX